Instructor's M
for

Modern
Carpentry

Building Construction Details
in Easy-to-Understand Form

by

WILLIS H. WAGNER
Professor Emeritus, Industrial Technology
University of Northern Iowa, Cedar Falls

HOWARD BUD SMITH
Author/Chief Editor
Lee Howard Associates
Bayfield, Wisconsin

and

MICHAEL B. KOPF
Technical Writer/Editor

Publisher
THE GOODHEART-WILLCOX COMPANY, INC.
Tinley Park, Illinois

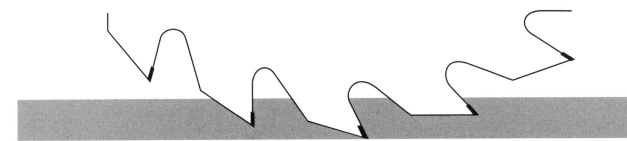

Contents

Introduction

Modern Carpentry provides up-to-date information about equipment, methods, and materials used in residential and light commercial construction. Although most of the examples are related to residential construction, the procedures and techniques detailed in this text may also be applied to a wide range of other frame buildings. The text is designed to provide fundamental information and instruction; however, the content has sufficient depth that it can be effectively used in advanced courses. This text addresses an industry need for persons who are able to understand and apply their skills and techniques.

Modern Carpentry also serves as an introduction to other building trades. A new section on mechanical systems presents technical information and "know-how" on house wiring, plumbing, and heating, ventilation, and air conditioning (HVAC). While these systems are normally installed by other tradespeople, a knowledge of the processes employed is helpful to the carpenter or home owner.

Each topic is addressed in easy-to-understand language. Tasks that require a series of steps to complete are set off from the text with a drawing of a sawhorse and the title of the task, such as:

 Constructing a Wall Section

Many of the more than 1600 illustrations are presented in color. Clear and concise drawings illustrate current carpentry techniques and standards, and are consistent with the skills required by today's carpenters.

Modern Carpentry places emphasis on the importance of safety on the worksite–whether it is on a job, at school, or at home. Students, do-it-yourselfers, and even experienced carpenters, must be able to recognize and correct unsafe conditions and practices. Unit 2, *General Safety Rules,* gives basic safety procedures and rules for the different topics covered in the text. Safety-related items throughout the text have been printed in red for emphasis.

USING THE TEXTBOOK

The units in **Modern Carpentry** are arranged in sequence with the phases of construction performed on a typical structure. Student ability, available facilities, and time constraints often determine the number of units your class will be able to cover. You may choose to include only those units (or series of units) best suited for a particular course. As each unit is complete in itself, the content flow is not disturbed.

The units on hand and power tools place emphasis on items most commonly used in building construction. Space limitations permit only brief descriptions and explanations of the tools. It is assumed that students using this textbook will have previously been enrolled in woodworking classes where the use of hand and power tools was adequately covered. If your students are lacking adequate knowledge of carpentry tools and equipment, **Modern Woodworking,** by Wagner and Kicklighter, may be used for review.

An outstanding feature of the **Modern Carpentry** textbook is the description of complicated processes in easy-to-understand terms. Explanations are clarified with carefully selected photographs, diagrams, and drawings. The textbook's detailed descriptions of construction processes makes it an excellent out-of-classroom resource. Completing out-of-classroom reading assignments will prepare students for in-class discussions. Students will also be able to better comprehend additional content presented through discussions, lectures, demonstrations, and other class-room resources.

Out-of-class study is especially important in vocational courses where students spend most of their in-class time working on construction activities. It not only saves valuable classroom time, but also provides students with the experience of researching and securing answers to questions. This out-of-class experience will also minimize the tendency of some students to depend on the instructor for direct assistance and help in finding solutions to every problem.

The textbook is divided into seven sections and 30 units. Section 1—Preparing to Build, includes information that needs to be mastered before beginning any carpentry project. Topics covered in Section 1 include building materials, wood identification, safety rules, hand tools, power tools, leveling instruments, and plans, specifications, and codes. Section 2—Footings, Foundations, and Framing, details topics involved in the preliminary stages of construction. Section 3—Closing In, deals with installation of shingles, windows, exterior doors and siding.

Section 4—Finishing, discusses such topics as insulation, wall and ceiling finishing, floor finishing, stair building, trim, and cabinetmaking. Section 5—Special Construction, involves topics such as chimneys, fireplaces, solar construction, remodeling, and painting. Section 6—Mechanical Systems, provides basic information and methods for installing wiring, plumbing, and HVAC. Section 7—Scaffolds and Careers, contains technical information on safe scaffolding, ladders, and an insight into career opportunities in carpentry.

Each unit is divided into several areas, including the introduction, text body, Important Terms, Test Your Knowledge questions, and Outside Assignments. A Math Review section, Technical Information section, Glossary, and Index are also included for easy reference.

Introduction

Unit topics are introduced in the first few opening paragraphs of each unit. In some cases, the introduction relates the new information to material that was discussed in the previous unit. Heighten the students' interest in the unit by having them read the introduction in class, and then asking them to list topics they feel might be included in the unit.

Text Body

New terms are printed in italics and are listed at the end of each unit. They are also included as entries in the index and defined in the glossary.

There are many trade tips integrated throughout the body of the text. These are identified with small icons representative of the topic being discussed. In many cases, the trade tips reinforce concepts presented in the text. In other cases, they provide additional information about a topic discussed in the text body.

For example:

 Steel components may have sharp edges. Use care in handling and wear leather gloves. Also, welded joints are hot and care should be exercised. Welded joints should be allowed to cool before handling.

End-of-Unit Materials

At the end of each unit there is a list of *Important Terms*, *Test Your Knowledge* questions, and *Outside Assignments*. The list of *Important Terms* and *Test Your Knowledge* questions may be used as either in-class or out-of-class assignments. The list of terms may be used as an oral or written quiz. Students can use the variety of true-false, short answer, and fill-in-the blank questions as a review of the unit.

Research and development in building construction have resulted in a tremendous number of new and improved materials and assembly systems. It is impractical, if not impossible, to adequately cover the field of carpentry in a single textbook. Thus, it is important that students spend some of their out-of-class time studying reference books, trade magazines, and manufacturers' literature.

A number of *Outside Assignments* are provided at the close of each unit in the textbook. They vary in complexity and form to match varying student interests and provide opportunities for using higher-order thinking skills.

Outside Assignments provide students with an opportunity to extend and enrich their learning experiences. Although completion of the assignments is not vital to accomplishing the objectives, it allows students to further their knowledge of the carpentry field.

Outside Assignments may be adapted to either individual or group work. Some activities are "paper-and-pencil" types while others require "hands-on." *Outside Assignments* are *not* designed to replace the regular construction activities being carried out in the laboratory. However, when the use of tools and materials is required, it may be appropriate to provide laboratory time to complete the projects.

To maintain some measure of control over the *Outside Assignments*, have students submit a brief written statement describing the problem and procedure they plan to follow. This is especially important for projects that require materials and equipment. At this point, you can assist the student(s) in finalizing their plans and make it possible to carry out the project in a practical and safe manner. Have students set the projects up as though they are "doing a job." Instruct them to keep records including a bill of materials, billable work hours, injury reports, and any other applicable records. These "records" and a written summary should be included in the student notebooks for evaluation at the close of the term.

Appendix A
Carpentry Math Review

An understanding of basic mathematics is a tool a carpenter will find as useful as any saw, drill, or hammer. Math is used to ensure that roofs slope properly, stairs rise evenly, and studs fit correctly. Appendix A, *Carpentry Math Review,* provides a "refresher course" covering the following topics:

- The use of rules, tapes, and squares.
- Working with fractions and decimals.
- Knowing and using formulas.
- Volume.

Appendix B
Technical Information

This part of the text includes technical information and specifications used in the carpentry trade. The following items are included:

- Standard abbreviations.
- Metrics in construction.
- Measurement conversions.
- Material specifications and grading information.
- Fastener specifications.
- Span tables.
- Truss types.
- Construction details.
- Insulation specifications and material types.
- Pipe sizes and specifications.
- Electrical specifications.

Glossary

The glossary provides detailed descriptions of technical terms commonly used in the trade. Although many of these terms are defined in the text, the glossary provides a convenient reference.

Index

Modern Carpentry contains a comprehensive index of all the concepts and topics presented in the text. The index allows the student to find information on a specific subject quickly and easily.

USING THE WORKBOOK

The Workbook for Modern Carpentry is designed to measure student achievement and comprehension of the material presented in the unit. The workbook reinforces the concepts and techniques described in the text. By using a variety of questions, you are able to fairly evaluate the students' progress. For the student, the workbook is a study guide and reference tool. The students can keep track of their own progress by using the chart on the last page of the workbook to list their scores and mark the activities completed.

The workbook has been organized to correspond with the content of the textbook. Objectives, questions, and problems in the workbook are listed in the same sequence as the information in the textbook.

When using the Workbook as a study guide, the students should first read and study the assigned material in the text. Then, without reference to the text, they should attempt to complete the Workbook assignment. Following completion of the assignment, the students may check their answers and work out scores.

You will find that the workbook provides an efficient method of checking on the students' progress. Selected items from the workbook can be used to develop periodic or final examinations. Informing the students that this procedure will be followed will likely stimulate their efforts to fully understand the information in the textbook rather than simply copying answers.

Answers to the questions and problems consist of words, letters, numbers, and simple drawings. Instruct students that words should be spelled correctly and letters and numbers should be carefully formed. It is highly recommended that the letters and words be printed. Stress to students that most tradespeople follow this same practice since the information will be easier to read and the possibility of errors greatly reduced.

Estimating amounts of material required for a building is an important part of carpentry. Procedures and examples are covered in various sections of the textbook, and estimating problems are included in the workbook. Extra space has been provided for making the calculations. Students should organize their figures so that they can check back through their work. This will also make it easier for you to evaluate the students' procedure.

Numbers should be "rounded out" and calculations "factored" as shown in the sample problems on pages 175 and 176, and other sections of the textbook. Simple diagrams and sketches are often helpful when solving complicated problems. These could be made on a separate sheet of paper or included in the space provided for calculations in the workbook.

A bank of computerized test questions (Goodheart-Willcox Test Creation Software) is also available. Instructors with access to a computer can create new tests using a variety of selection options.

USING THE INSTRUCTOR'S MANUAL

The **Instructor's Manual** is a useful tool for teaching carpentry using the **Modern Carpentry** textbook. The Instructor's Manual facilitates the development of a carpentry curriculum by offering suggestions on course content and providing resource material for the novice, as well as the experienced instructor. The Instructor's Manual details methods of implementing a carpentry curriculum in your own classroom. The unit resources provide detailed information regarding the instructional process for the course. In some units, procedure checklists are provided for evaluating students' psychomotor skills. A quiz is also included for each unit.

Answers to all questions in the text and workbook are also provided, along with answers to unit quizzes and section exams. Reproducible Masters, used to reinforce major concepts, are included in most units of the Instructor's Manual.

Developing a Carpentry Course

It is not possible for a publisher or author to provide a detailed course of study that will be suitable for every teaching situation. Some schools require that a specific program be followed. Course length, depending on local and/or state curriculum guides, can vary in length from 6 to 36 weeks. Classes may meet 1 to 5 times a week.

Only *you* can determine what material will best serve *your* students and how that material should be presented. Only *you* know the abilities of your students, the facilities and materials at your disposal, and the time available. **Modern Carpentry** is designed as a framework for a one-year course, but it may be condensed into one-semester. However, it is not advisable because of the depth of content.

There are different viewpoints on how to integrate carpentry into the high school vocational curriculum. Some schools choose to use carpentry as a capstone course. Most often, however, a carpentry course follows a two-semester course in woodworking fundamentals. This core course generally emphasizes the proper use of hand tools, and portable and stationary power equipment. After this carpentry course, students might pursue additional vocational training in an apprenticeship training program or other recognized training program.

Advisory Council

An advisory council is a committee of construction industry experts that advises the school department on course content, trends, and local labor needs. Invite members from a number of different fields. This panel helps you, as an instructor, to keep up-to-date and identify the requirements of a carpentry program.

An advisory council consists of members who are representative of the industry. For a carpentry

program, an advisory council might be formed with several general contractors (both commercial and residential) and carpenters. The course curriculum is based on their knowledge of the industry and your knowledge of teaching methodology and curriculum. The combination should allow you to present practical, up-to-date information to the students.

Course Content

Every carpentry course should follow the overall goals and objectives that have been established for the curriculum. These, in turn, are determined by expected student outcomes, labor needs, and career applications. Only when goals and objectives have been set should you begin to design a course.

There are more teaching curricula and materials available than most instructors could review. The board of education in many states and organizations such as the Associated General Contractors, have funded curriculum projects. Check with your regional and state boards of education, state department of vocational and technical education, or other related agencies.

Teaching Carpentry

Teaching carpentry, like many vocational subjects, requires a multifaceted approach to be successful. The combination of lectures, demonstrations, and other related activities may pose special problems for an instructor. Several factors must be considered when designing a carpentry course, including:

- Related academic skills.
- Varying student abilities.
- Budget limitations.
- Safety and liability.

Teaching Methods

Learning does not take place until the pupil wants to learn. Getting your students interested in learning is one of the most difficult parts of teaching. Motivation, therefore, is the first step in effective teaching. You need to stimulate your students so they want to learn.

Decide how you want to motivate your class. Make it interesting! You can use human curiosity or the competitive nature of people to your advantage. If you peak their curiosity, most students will be interested in seeing, hearing, and learning about new and different things. Challenge your students. Create an atmosphere where students will want to advance themselves. Make some projects a competition between groups.

Reward your students for a job well done. A word or two of praise can often change a person or class into success seekers. Emphasize a reward or personal gain. A reward could be something as simple as a free period or a field trip. Stress to your students that with true effort on their part and the proper training, the opportunity to secure a better job with greater earnings is much more likely to present itself.

Above all, you must be interested in what you are doing. Lack of interest on your part will convey itself to your students.

One of the quickest ways to lose students' interest is to be unprepared and present a meandering lesson. For effective teaching, you must carefully prepare the material, the learning environment, and the students.

Prepare yourself by becoming familiar with the material and having all the necessary teaching aids readily available and in good working condition. A variety of teaching methods and materials should be used to make topics clear to all students. Materials and methods that may be used include textbooks, workbooks, videos, lectures, demonstrations, overhead transparencies, interactive group discussions, hands-on activities, and field trips.

When Using
Posters or Blackboards

1. Use a pointer
2. Stand to one side
3. Face and talk to the class

The textbook is still the most important part of an instructional program. **Modern Carpentry** includes detailed explanations of construction procedures and materials and over a thousand illustrations. The text offers Review Questions, listings of New Terms, a variety of Outside Assignments, Step-by-step procedures, and a comprehensive Glossary. Your curriculum complements the text, and the text supports your teaching.

Short lectures supported with visual aids work best for conveying a large amount of information in a short time. Avoid lecturing for an entire class session. Student comprehension and attention decrease drastically after a certain point. Vary the teaching methods to add variety to the class. The unit resources outline various teaching methods that can be used in the classroom.

Balance the instruction by combining lectures with exercises and hands-on activities. Students need hands-on activities to apply what they have learned. Get your students involved! With available equipment, make videos of standard techniques and procedures. These can then be used for future classes. Each following class then can improve the video. Encourage students to solve problems on their own, using textbooks and other available resources.

Safety. A feature of the textbook is the special attention devoted to safety. Unit 2, General Safety Rules, covers the basic rules and regulations for construction work. At appropriate intervals throughout the text, safety considerations are included with instructions for performing specific operations. Safety-related items are printed in red for emphasis.

As the instructor of the course, you should continually emphasize the importance of safety in construction. Try to include further explanations related to the various safety rules and regulations as they apply to a specific situation. Students are more likely to develop a responsible attitude toward safety and follow recommended practices when they understand the reasoning behind the rules and regulations.

Construction Activities. The carpentry student should have the opportunity to perform as many construction operations as practical. In an ideal situation, the advanced student would have the opportunity to participate in work on a regular construction site.

In many vocational schools, "project houses" are constructed over a two- to three-year time span. Initially, financing problems may be encountered. The initial costs of the property and materials may seem tremendous. However, when the house is sold, the profit can be reinvested into a new project house for the following years.

In many cases, the site preparation and foundation work for a project house are completed by a qualified contractor. This allows the students to begin with the rough framing of the structure. Small work groups (3-6 students) should be formed to work as teams. Group the students and reassign them, as necessary, a few weeks into the project.

If your program cannot afford to finance a project house, research other possibilities. There are many government programs and not-for-profit groups that need help. Use your library resources and make contact with these programs. Offer the assistance of your student workers in exchange for a valuable learning experience.

When it is not possible to work on a "project house," full-scale construction of small buildings can often be carried out in the school shop and/or an outside area adjacent to the shop. Get the students involved in the community. Contact your local municipalities and once again, offer the assistance of your student workers.

Modern construction uses a broad range of component structures and prefabricated units—some of which may be built in the school shop and then removed for outdoor assembly or erection. Wall panels, roof trusses, stairs, and kitchen cabinet units are examples of this type of construction. Completed units may be disassembled and some of the materials reclaimed for other work, or they could be designed and built for use in a structure being erected by the class or a local contractor.

Considerable experiences can be gained by the student through the construction of scale models. Since this work is time-consuming, it may be best to erect only a wing or portion of a given structure. Since the model may be viewed from any angle, it shows how the finished structure will look in all three dimensions. This stimulates and motivates the students throughout the activity.

For model framing, the scale should be large enough to make work practical (min. 1 1/2″ = 1′-0″). Cut the framing members to match the nominal size of lumber, not the dressed dimensions. Use white pine, sugar pine, basswood, or other soft-textured woods that are free of heavy grain. Since the strips are usually too small for regular machine surfacing operations, rip them to size with a hollow-ground combination blade or with one of the new thin-kerf carbide-tipped blades.

In addition to model framing, various sections of a structure can be built using a larger scale (quarter-size, half-size, or even full-size). These constructions might consist of sill sections, cornice sections, rake sections; or such inside work as door jambs, stairs, finished flooring, and wall sections.

Models of the various component structures previously mentioned could be fabricated at a scale similar to framing structures. Also, special experimental designs of stressed-skin panels or box beams might serve as visual aids or as specimens that could be subjected to certain controlled testing.

Samples and Specimens. Throughout a building construction course, samples of modern materials should be available for the student to examine. Materials that are commonly used in a given area are easily obtained from local contractors and building supply dealers. Usable samples can also be obtained from most construction sites. (Obtain permission from the contractor before entering the site.) A partial list of

samples would include: species of lumber used locally; brick, stone, and concrete block; concrete forms; metal studs; sheathing lumber and boards; insulation materials; various types of shingles; finished flooring; ceiling tile; and inside molding and trim.

When storage space is available, various millwork items might be obtained. For example, a standard double-hung or other type of window unit, mounted upright on a base to permit operation and inspection, may be especially valuable. Also, sections of prefabricated kitchen cabinets or other building units would provide examples of design, methods of construction, and appropriate material.

Teaching Carpentry with Limited Facilities

Squeezing needed equipment out of a lackluster budget is not easy. If budget constraints pose a problem, try to improvise. Use the best-possible means to get your message across to the students. In some cases, the shear abundance of students can be addressed using one of the two following methods.

Pairing Students. After demonstrating a procedure or technique, pair more advanced students with beginners for *hands-on* activities to reinforce the concept(s) previously presented. The advanced student has the knowledge and initiative to work through the activity, and then act as a teacher. This person should answer questions and give training to his/her partner. The help should not extend too far beyond answering questions and demonstrating some techniques. The advanced student should not do the work for the partner. A disadvantage may occur when a behavioral problem student inhibits the study of his/her partner. If necessary, remedy this problem by pairing different partners at different times during the course.

Rotating Students. In many cases, the amount of time required to complete a certain phase of the project allows you to rotate students into particular tasks. When roofing a house, for example, allow one group of students to complete the first few courses and then rotate other students into their positions. However, rotate only one or two students at a time. This allows the "experienced" roofers to become the teachers.

Student Outcomes

Knowledge is the most important tool you can provide to your students. You must form a solid base in the fundamentals in order to build your students carpentry skills. Emphasize the importance of developing reading, writing, mathematics, and communication skills. Stress that, in addition to carpentry skills, employers often list the abilities to follow proper procedures, work with others, and communicate as the deciding factor among job applicants.

Varying Student Abilities

Students entering your classroom will undoubtedly possess a variety of skills. Some students may have extensive exposure to the carpentry trade. Some may possess basic woodworking skills, while others will not. Students will also have a wide variety of career goals. Some students take a carpentry course for exploratory purposes, while others intend to use the skills throughout their careers.

Students entering your classroom also possess a wide range of potential abilities. In many high school programs, special needs students are regularly mainstreamed into vocational-technical programs. Special needs students generally include mentally or physically impaired students. Gifted students may also enroll in technical courses.

Identifying the Special Needs Student. Ideally, special needs students will be identified by the school psychologist. The instructor is then informed of the situation. If the special needs students have not been identified, you must assume the responsibility of doing so. Once the students are identified, specialized learning programs and/or teaching methods may be implemented.

Characteristics of special needs students are included in the following list. An infrequent occurrence does not denote that this is a special needs student.

- ❏ Inability to communicate effectively (reading, writing, oral expression, and questioning).
- ❏ Short attention span or lack of interest.
- ❏ Frequent disruptive behavior.
- ❏ Lack of motivation.
- ❏ Poor computational skills.
- ❏ High rate of absenteeism.
- ❏ Impaired speech, hearing, and/or sight.
- ❏ Lack of self-confidence.

Mainstreaming. In the past, most special needs students were segregated into separate rooms for instruction. The rooms were generally staffed with instructors trained to work with special needs students. The instructors, however, were not well trained in vocational-technical disciplines. Mainstreaming is a popular method of integrating special needs students into vocational-technical programs.

A variety of special needs students are commonly found in any classroom. Some may have limited mobility; others may have vision or hearing problems. Always remember that a special needs student needs to be challenged just like any other student. The type of challenge should be based on the students' potential. You may need to modify your teaching methods, allowing the special needs students to reach their goals

A vision-impaired student should be allowed to explore the construction site with a student companion. The exploration will allow the student to become familiar with locations of equipment and other items. In the classroom, seat the student toward the front of the class. It is not necessary, however, to seat the student in the first row. When lecturing, be specific in your references. Pay close attention to the needs of the student.

Speak in a normal, but distinct, tone of voice when working with a hearing-impaired student. Pronounce words clearly to allow the student full comprehension of the message. Face the students in the classroom when lecturing or entering in a discussion. This allows your voice to be heard more clearly.

Identifying the Gifted Student. The gifted student may be more difficult to identify and just as challenging to instruct as the special needs student. In some cases, an instructor is informed of a gifted student by a counselor. In many cases, however, gifted students have not been identified. Vocational instructors should be familiar with gifted students' characteristics and modify teaching methods accordingly. Common characteristics of gifted students include:

- ❏ When properly motivated, displays more interest and a longer attention span.
- ❏ May be more inclined to question or comment on the subject.
- ❏ May seem restless, bored, or disinterested.
- ❏ Consistently completing work ahead of others.
- ❏ Consistently scores high on exams, etc.
- ❏ Pays close attention to detail on a procedure or technique that is generally taken for granted.

A gifted student should be constantly challenged to maintain interest in the subject matter. Assign tasks with greater complexity, such as calculating rafter angles. A gifted student might also be used as a classroom assistant. He/she might be able to estimate materials that are needed for upcoming projects or other similar activities.

Additional Student Experiences and Resources

A variety of supplemental materials may be used to allow the students to further develop their interest in carpentry. Not all learning occurs in the classroom.

Students should be encouraged to read supplemental material and participate in field trips. In either case, the activity should be followed up by some means of discussion and/or evaluation.

Outside Assignments. Students should thoroughly study assigned textbook units. Time may be given in class for review or questions but the reading and *Review Questions* should be completed on the student's own time. Stress to the students that the technical and instructional information must be studied thoroughly for complete comprehension.

Extensive programs of research and development in all areas of building construction have resulted in an "explosion of knowledge," this makes it impractical to cover the field of carpentry in a single textbook. Thus, it is essential that students spend some of their out-of-class time in the study of reference books, trade magazines, and manufacturers' literature. A number of suggested activities that may provide topics and ideas for outside assignments of this nature are included at the end of each unit.

Field Trips. Teachers of building construction classes have found field trips to be highly effective. Every community, large or small, offers opportunities in this area. Field trips should be carefully planned and arrangements made ahead of the actual visit. More than one trip can be made to a particular project. Several trips in close sequence during rough framing aspects may be followed by additional visits, spaced further apart, during finishing stages. The greatest instructional gains, of course, can be made when the work in progress at the construction site is in phase with the content being covered in regular class sessions.

Students should receive a briefing prior to the trip. They will then know what to look for and can better understand what they observe. If possible, have a representative from the project visit your classroom and give a presentation. A follow-up evaluation and discussion session is extremely valuable, especially when it can be held shortly after returning. You might choose to have students write a brief report on the field trip.

Cooperative Training. A cooperative training program, or internship, is an agreement between a school and a local contractor. This agreement allows a student to work part time for a contractor. The program is designed to help mature, advanced students pursue career goals. The student earns money while gaining valuable experience. The employer gains a good worker at minimum cost. In addition, that student may move into the company once his/her studies are finished. Assist students in determining their goals and

objectives so they can look for a job that holds the same focus.

Organizations. Encourage students to get involved in recognized organizations such as the Vocational Industrial Clubs of America (VICA). These organizations promote vocational-technical education and provide worthwhile experiences for the students. Organizations also help to develop interpersonal communication skills. A person, regardless of his or her career, must be able to communicate effectively with coworkers.

Evaluation

Students learn in a variety of ways. Many are very visual; others learn better from lectures and demonstrations. In a similar manner, a variety of techniques must be used to fairly evaluate student progress. The Instructor's Manual for **Modern Carpentry** provides several means of evaluating students in a carpentry course.

In a broad view of evaluation, give students credit for work well done. Provide frequent reports showing each student's accomplishments. Hold the students responsible for their progress and offer motivation to reach their goals. Adjust your teaching methods and emphasis as needed.

Unit Quizzes

The unit resources include a short quiz for each unit. The quiz should be used to verify the students comprehension of the information presented in the unit. You may also use the quizzes to identify any shortcomings of the instruction. You can record the class results of each question. Questions that are often incorrectly answered may denote shortcomings of the instruction on a particular subject. You may use the results to modify the instruction when the unit is presented again.

Section Exams

Seven section exams provide overall evaluation of the material contained in the respective units. See the table of contents for a detailed list of units included in each section. A variety of question types are used including True-False, Multiple Choice, Identification, Matching, Completion, and Short Answer.

Procedure Checklists

Procedure checklists are used to evaluate common procedures or techniques. Productivity plays a major role on the job and should be evaluated in the classroom as well. For information regarding the use of a procedure checklist, see the "Procedure Checklist" heading described under "UNIT RESOURCES."

Employability Skills

Employability skills are a vital part of industrial technology/vocational education programs. They should be included in any occupation-focused course.

Identifying Employment Opportunities
- Identify requirements for job.
- Investigate educational opportunities.
- Investigate occupational opportunities.
- Locate resources for finding employment.
- Confer with prospective employers.
- Identify job trends.

Applying Employment Seeking Skills
- Locate job openings.
- Document skills and abilities (resume).
- Prepare for interview.
- Participate in interview.
- Complete required forms.
- Write application letter.
- Write follow-up letter.
- Evaluate job offer.
- Evaluate job rejection.

Interpreting Employment Capabilities
- Match interest to job area.
- Match aptitudes to job area.
- Verify abilities.
- Identify immediate work goal.
- Develop career plan.

Demonstrating Appropriate Work Behavior
- Exhibit dependability.
- Demonstrate punctuality.
- Follow rules and regulations.
- Recognize the consequences of dishonesty.
- Control emotions.
- Assume responsibility for their own decisions and actions.
- Exhibit pride and loyalty.
- Exhibit ability to handle pressures and stress.
- Demonstrate ability to set priorities.
- Demonstrate problem-solving skills.

Maintaining Safe and Healthy Environment
- Comply with safety and health rules.
- Select correct tools and equipment.
- Use equipment correctly.
- Use personal protective equipment properly.
- Use appropriate action during emergencies.
- Maintain clean and orderly work area.
- Demonstrate personal hygiene and cleanliness.

Maintaining Businesslike Image
- Participate in company or agency orientation.
- Demonstrate knowledge of company or agency products and services.
- Exhibit positive behavior.
- Read current job-related publications.
- Support and promote employer's company image and purpose.
- Maintain appearance to comply with company standards.

Maintaining Working Relationships with Others
- Work productively with others.
- Show empathy, respect, and support for others.
- Demonstrate procedures and assist others when necessary.
- Recognize, analyze, and solve or refer problems.
- Minimize occurrence of problems.
- Channel emotional reaction constructively.

Communicating on the Job
- Read and comprehend written communications and information.
- Speak effectively with others.
- Use job-related terminology.
- Listen attentively.
- Write legibly.
- Follow written and oral directions.
- Ask questions.
- Locate information in order to accomplish tasks.

Adapting to Change
- Recognize need to change.
- Demonstrate willingness to learn.
- Demonstrate flexibility.
- Participate in continuing education.
- Seek worthy challenges.
- Adjust career goals/plan as needed.

Understanding How a Business Works
- Recognize the role of business in the American enterprise system.
- Identify general responsibilities of employees.
- Investigate opportunities and options for business ownership.
- Identify the planning processes needed to open a business.
- Participate in meetings.

Supplemental Reading and Professional Resources

An in-house library of current magazines, catalogs, journals, brochures, users manuals, advertisements, and other informative publications is an excellent investment. Many publications are free for the asking. The task of acquiring literature can be given as student projects.

In addition to current periodicals, have other trade references pertaining to the carpentry field on hand. This might include OSHA manuals, *Architectural Graphics Standards, Machinery's Handbook,* and texts on carpentry, construction, manufacturing, materials and processes, welding, electronics, etc. See the section titled, *"Resources for Carpentry Education"* for popular periodicals.

Resources for Carpentry Education

Various resources are available for additional carpentry information. The following list of periodicals provide information pertaining to the carpentry trade, and woodworking in general.

Builder
The National Association of Home Builders
655 15th Street, NW
Suite 475
Washington, DC 20005

Builder & Contractor
Associated Builders & Contractors, Inc.
729 15th Street, NW
Washington, DC 20005

Building Design & Construction
Cahners Publishing Company
275 Washington Street
Newton, MA 02158-1630

Computers for Design & Construction
310 E. 44th Street
New York, NY 10017

Construction Digest
Construction Digest, Inc.
7355 Woodland Drive
Indianapolis, IN 46278

Construction Equipment
Cohners Plaza
1350 E. Touhy Avenue
P.O. Box 5080
Des Plaines, IL 60017-5080

Constructor
Associated General Contractors of America
1957 East Street, NW
Washington, D.C. 20006

Exteriors
Edgell Communications, Inc.
7500 Old Oak Boulevard
Cleveland, OH 44130

Fine Homebuilding
The Taunton Press
63 South Main Street
Box 355
Newton, CT 06470-9989

Fine Woodworking
The Taunton Press
63 South Main Street
Box 355
Newton, CT 06470-9989

Historic Preservation
1785 Massachusetts Avenue, NW
Washington, DC 20036

Interior Construction
104 Wilmot Road, Suite 201
Deerfield, IL 60015-5195

Interior Design
A Cohners Publication
P.O. Box 1970
Marion, OH 43305

The Journal of Light Construction
P.O. Box 689
Mt. Morris IL 61054

Residential Construction Guidelines
   American Iron land Steel Institute
1101 17th St. NW
Washington DC 20036-4700

Journal of Research of the National
   Bureau of Standards
U.S. Government Printing Office
Washington, DC 20402

Multi-Housing News
Gralla Publications
1515 Broadway
New York, NY 10036

Nation's Building News
The National Association of Home Builders
15th and M Streets, NW
Washington, DC 20005

Practical Homeowner
Practical Homeowner Publishing Co.
825 Seventh Avenue
New York, NY 10019

Professional Builder
Reed Publishing USA
275 Washington Street
Newton, MA 02158-1630

Progressive Architecture
Penton Publishing
600 Summer Street
P.O. Box 1361
Stanford, CT 06904

Qualified Remodeler
20 E. Jackson Boulevard
Chicago, IL 60604

Wood and Wood Products
Vance Publishing Corporation
400 Knightsbridge Parkway
Lincolnshire, IL 60069

Woodshop News
Pratt Street
Essex, CT 06426

Woodsmith
2200 Grand Avenue
Des Moines, IA 50312

A valuable resource for building construction courses lies in manufacturers' and suppliers' catalogs and various forms of descriptive literature they distribute. Some of this material is available from local building supply dealers while other items may be secured by writing directly to the company. Single copies are usually available free of charge when the letter of request is written on official school stationery.

A word of caution: do not secure these materials unless you plan to make good use of them. Some companies have stopped making their literature available because it is often wasted. In your letter, assure the organization that the materials they send will be effectively used.

Trade associations are formed when a number of manufacturers in a related field join together and pool their efforts in such areas as research and development, product standards and inspection, public information, and sales promotion. These associations usually have a wide range of literature. In an initial letter, you should simply request a listing of the items they have available and would be willing to furnish. When this list is received, requests for specific items can be made.

Several companies and trade associations make a selection of 16 mm films and videotapes available for school use. They show the fabrication of various building materials and products. Most of them can be obtained free of charge. The borrower pays only the return postage.

The following is a selected listing of building material manufacturers' associations and related groups:

Access for the Handicapped
5014 42nd Street, NW
Washington, DC 20016

ASA Acoustical Society of America
500 Sunnyside Boulevard
Woodberry, NY 11797

ASC Adhesive and Sealant Council, Inc.
1500 Wilson Boulevard, Suite 515
Arlington, VA 22209-2495

ACEA Allied Construction Employers Association
180 N. Executive Drive
Brookfield, WI 53008

AA Aluminum Association
750 Third Avenue
New York, NY 10017

AAA American Arbitration Association
140 W. 51st Street
New York, NY 10020

**AASHTO American Association of State Highway
and Transportation Officials**
444 N. Capitol Street, NW
Suite 225
Washington, DC 20001

ABCA American Building Contractors Association
11100 Valley Boulevard, Suite 120
El Monte, CA 91731

ACI American Concrete Institute
22400 W. Seven Mile Road
Detroit, MI 48219

AHA American Hardboard Association
520 N. Hicks Road
Palatine, IL 60067

**AHMU American Hardware Manufacturers
    Association**
931 N. Plum Grove Road
Schaumburg, IL 60173

AHLI American Home Lighting Institute
435 N. Michigan Avenue,
Suite 1717
Chicago, IL 60611

└ **AIA American Institute of Architects**
1735 New York Avenue, NW
Washington, DC 20006

└ **AIC American Institute of Constructors**
20 S. Front Street
Columbus, OH 43215

└ **AITC American Institute of Timber Construction**
11818 S.E. Mill Plain Boulevard
Vancouver, WA 98684

└ **ALSC American Lumber Standards Committee**
P.O. Box 210
Germantown, MD 20874

ANSI American National Standards Institute
1430 Broadway
New York, NY 10018

APA American Plywood Association
P.O. Box 11700
Tacoma, WA 98411

**ASTM American Society for Testing and
    Materials**
1916 Race Street
Philadelphia, PA 19103

ASID American Society of Interior Designers
1430 Broadway
New York, NY 10018

**ASPE American Society of
    Professional Estimators**
3617 Thousand Oaks Boulevard, Suite 210
Westlake, CA 91362

ASA American Subcontractors Association
1004 Duke Street
Alexandria, VA 22314

American Wood Preservers Bureau
P.O. Box 6085
Arlington, VA 22206

**AAMA Architectural Aluminum
    Manufacturers Association**
2700 River Road, Suite 118
Des Plaines, IL 60018

AWI Architectural Woodwork Institute
2310 S. Walter Reed Drive
Arlington, VA 22206

**AIANA Asbestos Information
    Association/North America**
1745 Jefferson Davis Hwy., Suite 509
Arlington, VA 22202

└ **ARMA Asphalt Roofing
    Manufacturers Association**
757 Third Avenue
New York, NY 10017

└ **ABC Associated Builders and Contractors, Inc.**
729 15th Street, NW
Washington, DC 20005

└ **AGC Associated General Contractors of America**
1957 East Street, NW
Washington, DC 20006

ASC Associated Specialty Contractors
7315 Wisconsin Avenue
Bethesda, MD 20814

**AWCI Association of the Wall and Ceiling
    Industries International**
25 K Street, NE
Suite 300
Washington, DC 20002

└ **BHMA Builder's Hardware Manufacturers
    Association, Inc.**
60 E. 42nd Street, Rm. 511
New York, NY 10165

**BOCA Building Officials and Code
Administrators International**
4051 W. Flossmoor Road
Country Club Hills, IL 60477

Building Materials Research Institute, Inc.
501 5th Avenue, #1402
New York, NY 10017

BRB Building Research Board
2101 Constitution Avenue, NW
Washington, DC 20418

BSC Building Systems Council
15th and M Street, NW
Washington, DC 20005

CRA California Redwood Association
591 Redwood Highway, Suite 3100
Mill Valley, CA 94941

Canadian Forest Products Ltd.
440 Canfor Avenue
New Westminster, B.C., Canada V3L 3C9

Canadian Wood Council
84 Albert Street
Ottawa, Canada K1P 6A4

CRI Carpet and Rug Institute
P.O. Box 2048
Dalton, GA 30722-2048

**CISCA Ceilings and Interior Systems
Construction Association**
104 Wilmot, Suite 201
Deerfield, IL 60015

CTI Ceramic Tile Institute
700 N. Virgil Avenue
Los Angeles, CA 90029

CRSI Concrete Reinforcing Steel Institute
933 N. Plum Grove Road
Schaumburg, IL 60195

**CIEA Construction Industry
Employers Association**
625 Ensminger Road
Tonawanda, NY 14150

**CIMA Construction Industry
Manufacturers Association**
111 E. Wisconsin Avenue, Suite 940
Milwaukee, WI 53202-4879

**CPMA Construction Products
Manufacturing Council**
P.O. Box 21008
Washington, DC 20009-0508

CSI Construction Specifications Institute
601 Madison Street
Alexandria, VA 22314

CABO Council of American Building Officials
5203 Leesburg Pike, Suite 708
Falls Church, VA 22041

DHI Door and Hardware Institute
7711 Old Springhouse Road
McLean, VA 22102-3474

EPA Environmental Protection Agency
401 M Street, SW
Washington, DC 20460

FTI Facing Tile Institute
P.O. Box 8880
Canton, OH 44711

FHA Federal Housing Administration
451 7th Street, SW
Room 3158
Washington, DC 20410

FPRS Forest Products Research Society
2801 Marshall Ct.
Madison, WI 53705

GBCA General Building Contractors Association
36 S. 18th Street
P.O. Box 15959
Philadelphia, PA 19103

GA Gypsum Association
1603 Orrington Avenue, Suite 1210
Evanston, IL 60201

**HPMA Hardwood Plywood
Manufacturers Association**
P.O. Box 2789
Reston, VA 22090

17

IFI Industrial Fasteners Institute
1505 E. Ohio Building
Cleveland, OH 44114

ICAA Insulation Contractors
  Association of America
15819 Crabbs Branch Way
Rockville, MD 20855

ICBO International Council of Building Officials
5360 S. Workman Mill Road
Whittier, CA 90601

IILP International Institute for Lath and Plaster
795 Raymond Avenue
St. Paul, MN 55114

LIUNA Laborers' International
  Union of North America
905 16th Street, NW
Washington, DC 20006-1765

Manufactured Housing Institute
1745 Jefferson Davis Highway, #511
Arlington, VA 22202

MFMA Maple Flooring
  Manufacturers Association
60 Revere Drive, Suite 500
Northbrook, IL 60062

MLSFA Metal Lath/Steel Framing Association
600 S. Federal, Suite 400
Chicago, IL 60605

Mineral Fiber Products Bureau
509 Madison Avenue
New York, NY 10022

NAFCD National Association of
  Floor Covering Distributors
13-126 Merchandise Mart
Chicago, IL 60654

NAHB National Association of Home Builders
15th and M Street, NW
Washington, DC 20005

NAHRO National Association of Housing
  Redevelopment Officials
1320 18th Street, NW
Washington, DC 20036

NARSC National Association of
  Reinforcing Steel Contractors
10382 Main Street
P.O. Box 225
Fairfax, VA 22030

NAWIC National Association of
  Women in Construction
327 S. Adams Street
Fort Worth, TX 76104

National Building Code
American Insurance Association
85 John Street
New York, NY 10038

NBMA National Building
  Manufacturers Association
142 Lexington Avenue
New York, NY 10016

National Building Material
  Distributors Association
1701 Lake Avenue, Suite 170
Glenview, IL 60025

NCSBCS National Conference of States on
  Building Codes and Standards
481 Carlisle Drive
Herndon, VA 22070

NCA National Constructors Association
1101 15th Street, NW, Suite 1000
Washington, DC 20005

NCRPM National Council on Radiation
  Protection and Measurement
7910 Woodmont Avenue, Suite 800
Bethesda, MD 20814

NFPA National Fire Protection Association
Batterymarch Park
Quincy, MA 02269

NFPA National Forest Products Association
1250 Connecticut Avenue, NW, Suite 200
Washington, DC 20036

Northern Hardwood and Pine
  Manufacturers Association
Suite 207 Northern Bldg.
Green Bay, WI 54301

National Housing Rehabilitation Association
1726 18th Street, NW
Washington, DC 20009

NKCA National Kitchen Cabinet Association
P.O. Box 6830
Falls Church, VA 22046

**NLBNMA National Lumber and Building
Material Dealers Association**
40 Ivy Street, SE
Washington, DC 20003

National Mineral Wool Association
382 Springfield Avenue
Summit, NJ 07901

**NOFMA National Oak Flooring
Manufacturers Association**
P.O. Box 3009
Memphis, TN 38173-0009

L **NPCA National Paint and Coatings Association**
1500 Rhode Island Avenue, NW
Washington, DC 20005

National Particleboard Association
2306 Perkins Place
Silver Spring, MD 20910

L **NRCA National Roofing Contractors Association**
1 O'Hare Center
6250 River Road
Rosemont, IL 60018

National Wood Window and Door Association
205 Touhy Avenue
Park Ridge, IL 60068

**NWMA National Woodwork
Manufacturers' Association**
400 W. Madison Street
Chicago, IL 60606

**Operative Plasterers' and Cement Masons'
International Association of the United States
and Canada**
1125 17th Street, NW, 6th Floor
Washington, DC 20036

PSIC Passive Solar Industries Council
2836 Duke Street
Alexandria, VA 22314

**RCSHSB Red Cedar Shingle and
Handsplit Shake Bureau**
515 116th Avenue, NE, Suite 275
Bellevue, WA 98004

**RFCA Resilient Flooring and
Carpet Association, Inc.**
14570 E. 14th Street, Suite 511
San Landro, CA 94578

**Scaffolding, Shoring, and Forming
Institute, Inc.**
1230 Keith Building
Cleveland, OH 44115

SMA Screen Manufacturers Association
655 Irving Park, Suite 201
Chicago, IL 60613-3198

SWI Sealant and Waterproofers Institute
3101 Broadway, Suite 300
Kansas City, MO 64111

**SIGMA Sealed Insulating Glass
Manufacturers Association**
111 E. Wacker Drive, Suite 600
Chicago, IL 60601

**SBCCI Southern Building Code Congress
International, Inc.**
900 Montclair Road
Birmingham, AL 35213

SFPA Southern Forest Products Association
Box 52468
New Orleans, LA 70152

SDI Steel Door Institute
712 Lakewood Center N
14600 Detroit Avenue
Cleveland, OH 44107

SJI Steel Joist Institute
1205 48th Avenue, N, Suite A
Myrtle Beach, SC 29577

SWI Steel Window Institute
1230 Keith Building
Cleveland, OH 44115

SMA Stucco Manufacturers Association
14006 Ventura Boulevard
Sherman Oaks, CA 91423

SBA Systems Builders Association
P.O. Box 117
West Milton, OH 45383

**TIMA Thermal Insulation
    Manufacturers Association**
29 Bank Street
Stanford, CT 06901

**TCAA Tile Contractors
    Association of America, Inc.**
112 N. Alfred Street
Alexandria, VA 22314

TCA Tile Council of America
P.O. Box 2222
Princeton, NJ 08542

Truss Plate Institute
583 D'Onofrio Drive, Suite 200
Madison, WI 53719

**UBC United Brotherhood of Carpenters and
    Joiners of America**
101 Constitution Avenue, NW
Washington, DC 20001

**U.S. Department of Labor/Occupational Safety
    and Health Administration**
200 Constitution Avenue, NW
Washington, DC 20210

U.S. Forest Products Laboratory
One Gifford Pinchot Drive
Madison, WI 53705-2398

**UURWAW United Union of Roofers,
    Waterproofers and Allied Workers**
1125 17th Street, NW, 5th Floor
Washington, DC 20036

Vinyl Siding Institute
355 Lexington Avenue
New York, NY 10017

WMA Wallcovering Manufacturers Association
355 Lexington Avenue
New York, NY 10017

WWPA Western Wood Products Association
Yeon Building
522 S.W. 5th Avenue
Portland, OR 97204

Wood Truss Council of America
111 Wacker Drive
Chicago, IL 60601

UNIT RESOURCES

Unit resources provide activities and teaching suggestions. They also provide answers to all questions, exercises, and activities in the textbook and workbook. Reproducible masters, procedure checklists (where applicable), and a unit quiz are also included.

An important feature of the unit resources is the "Trade-Related Math" section. Basic math concepts are presented in each unit and then related to text information.

Objectives

Objectives are listed for each of the 30 units. The goals presented involve basic concepts, skills, and understandings students should derive from their study. You will want to adapt these goals to your particular teaching situation and student composition. Preparation for instruction with these objectives in mind will lend direction to the use of **Modern Carpentry** in your course.

Instructional Materials

The instructional materials section lists all the resources provided for teaching the unit. Prepare the majority of the instructional materials prior to teaching the unit. This will allow you to review the material while you are duplicating it. Unit quizzes should not be copied until just prior to giving the evaluation. This will allow you to maintain confidentiality of the quiz.

Trade-Related Math

This section of the unit resources includes one or two basic math concepts that can be related to the text information. The math concepts can be integrated into the unit presentation, or used as a follow-up at the conclusion of the unit.

Instructional Concepts and Student Learning Experiences

This section provides teaching suggestions and strategies for using the textbook, workbook, and/or reproducible masters. Information and teaching suggestions should be tailored to fit the needs of the curriculum. Other activities and strategies can be developed from ideas that are presented in the unit resources. Maintain a file of ideas for future use.

Reinforce. It is generally accepted that the more ways a student is exposed to a given concept, the greater the chances for his/her understanding and retention of the material. A variety of learning experiences are designed to meet the reinforcement needs of students.

Extend. The teaching suggestions in the Instructor's Manual are directed to students at a variety of ability levels. You may choose some of the assignments to encourage highly motivated students to extend their learning experience outside of the classroom. These types of activities allows students to relate text information to other experiences, particularly life skills.

Enrich. Enrichment activities are designed to help students learn more about topics introduced in the text. These types of activities, such as research and survey activities, give students the opportunity to enrich their learning through more in-depth study.

Reteach. Studies have shown that students respond differently to different teaching methods and techniques. Therefore, these materials provide suggestions for several strategies that can be used to teach the concepts in the text. This allow you to choose a different strategy to reteach students who showed a low response to a previous strategy.

Safety

In applicable units, safe work practices are stressed. In the text, safety considerations are printed in red. The **Instructor's Manual** calls your attention to these safety considerations so they will not be overlooked in your presentation. Stress the need for adequate personal protective equipment during all shop work.

Answer Keys

The answers are included within the corresponding unit. Answers to section exams are included with the last unit in each section.

Reproducible Masters

Reproducible masters are designed to be used in several ways—duplicating masters, overhead transparencies, and test masters. Duplicating masters are designed to be copied and used by the students. Many duplicating masters are designed to be used along with specific transparency masters. You can make a transparency and project it while the students work on a copy of the duplicating master. The reproducible masters can also be projected using an opaque projector.

The reproducible masters can also be used for periodic tests and examinations. A test master can be prepared by masking out selected information and using a number or letter to identify the item.

Unit Quiz

True-false, multiple choice, and identification questions are used. Students should place their answers on the blank to the left of the question number where applicable. This facilitates grading the quizzes.

Section Exam

Section exams are included for each of the seven sections in the text. A variety of question types are included on each exam. Section exams should be given to students for evaluation after they have completed the textbook and workbook questions, and quizzes for all units in the section. Students should prepare for the section exam by reviewing the appropriate unit materials and quizzes. Answers for the section exams are included with the last unit of each section.

Procedure Checklist

As the instructor, you would fill out a procedure checklist as a student performs certain tasks. A range of scores (1-5) is listed for each measurable objective. You can circle the score appropriate for each task performed. Since the checklists are general in nature, you may modify the checklists to evaluate more specific tasks.

Some of the tasks involved in carpentry must be performed by several people working together. In these cases, write all students' names at the top of the procedure checklist and evaluate them as a group.

Goodheart-Willcox welcomes your input.

If you have comments, corrections, or suggestions regarding the textbook or its supplementaries, please send them to:

Managing Editor–Technology
Goodheart-Willcox Co., Incorporated
18604 West Creek Drive
Tinley Park, IL 60477

1

Building Materials

OBJECTIVES

Students will be able to:

- ☐ Describe the hardwood and softwood classifications.
- ☐ Define moisture content (M.C. and E.M.C.).
- ☐ Recognize and name common defects in lumber.
- ☐ Define lumber grading terms.
- ☐ Calculate lumber sizes according to established industry standards.
- ☐ Explain plywood, hardboard, and particleboard grades and uses.
- ☐ Identify nail types and sizing units.
- ☐ List precautions to observe while working with treated lumber.
- ☐ Recognize types of engineered lumber and list their uses and advantages.
- ☐ Discuss the uses of metal structural materials and suggest their advantages/disadvantages.
- ☐ Identify a variety of metal framing connectors and indicate where each is used.

INSTRUCTIONAL MATERIALS

Text: Pages 17-50
   Important Terms, page 46
   Test Your Knowledge, page 47
   Outside Assignments, page 47
Workbook: Pages 7-14
Instructor's Manual:
   Reproducible Master 1-1, *Basic House Parts*
   Reproducible Master 1-2, *Typical Grade Stamps*
   Reproducible Master 1-3, *Nail Types, Parts A and B*
   Reproducible Master 1-4, *Nail Sizes*
   Unit 1 Quiz

TRADE-RELATED MATH

The board foot is the standard unit of measure for lumber. One board foot is equivalent to a piece of lumber measuring 1″ thick by 12″ square. To calculate board feet, use the following formula:

$$\text{Bd. ft.} = \frac{\text{No. pcs.} \times T \times W \times L}{12}$$

INSTRUCTIONAL CONCEPTS AND STUDENT LEARNING EXPERIENCES

LUMBER

1. Identify types of materials that are considered (and not considered) to be lumber. Have different types of materials available for student inspection, and ask them to identify each.

WOOD STRUCTURE AND GROWTH

1. Using a cross-sectional piece of a log, ask students to identify the following parts and their purpose(s): heartwood, sapwood, pith, wood rays, annual rings, wood rays, cambium, and bark.
2. Discuss the growth of a tree, and how the parts must interact for proper growth.

KINDS OF WOOD

1. List the differences between hardwood and softwood. Stress that the physical "hardness" of a wood does not necessarily indicate whether a wood is classified as a hardwood or softwood.
2. Using the full-color wood samples on pages 48-50 of the text, have students identify the wood by color. Discuss other means for identifying wood, such as weight and material properties.
3. Using unlabeled wood specimens, have students identify various types of wood. Ask them to list the criteria that they used to determine the types of wood.

CUTTING METHODS

1. Discuss the methods of cutting lumber at a sawmill, including plain-sawed (softwoods) and

quarter-sawed (hardwoods). List the advantages and disadvantages of each type of cutting.

MOISTURE CONTENT AND SHRINKAGE

1. Describe the method used to determine the moisture content of wood.
2. If an oven is available, demonstrate the procedure for determining moisture content as outlined in the text. Use various specimens of wood to conduct the demonstration.
3. Have the students define the term "equilibrium moisture content." Cite examples where the relative humidity of the air affects the carpentry trade. Identify parts of the country where more or less humidity can be expected.
4. Using Reproducible Master 1-1, *Basic House Parts,* have students identify parts of the house where particular attention must be paid to the moisture content of lumber.

SEASONING LUMBER

1. Identify the primary means of seasoning lumber: air drying and kiln drying. Discuss the type of preparation necessary before lumber can be seasoned.
2. Review the procedure for determining moisture content by drying a sample in an oven.
3. Identify the two basic types of moisture meters. If possible, demonstrate the use of a moisture meter, and allow students to determine the moisture content of various wood samples.

LUMBER DEFECTS

1. Discuss the problems associated with wood defects in lumber, including loss of strength, durability, and usefulness.
2. Display several pieces of lumber that have various wood defects. Have students identify the defects, and list the causes of the defects. Stress the importance of using good-quality lumber on a construction project.

SOFTWOOD GRADES

1. Identify the principal organizations involved in grading lumber, including the American Lumber Standards Committee, Western Wood Products Association, Southern Forest Products Association, and California Redwood Association.
2. Using the charts in Figure 1-18, page 24, and Reproducible Master 1-2, *Typical Grade Stamps,* discuss the various grades of softwood lumber.

3. Obtain price lists from a local lumber yard, and compare the costs of the different grades of softwood lumber. Stress the importance of obtaining good-quality lumber for a construction project, but not specifying too good of a grade, and thus increasing construction costs.

HARDWOOD GRADES

1. Identify the three major classifications of hardwood lumber: FAS, Selects, and No. 1 Common. Obtain samples of the different grades and discuss the methods used to grade the lumber. Reproducible Master 1-2, *Typical Grade Stamps,* may also be used to enhance the discussion.

LUMBER STRESS VALUES

1. Discuss the means for assigning lumber stress values.

LUMBER SIZES

1. Review the charts in Figure 1-19, page 26, stressing the difference between nominal and dressed sizes. List the various nominal sizes of lumber and have students identify the respective dressed sizes.

FIGURING BOARD FOOTAGE

1. Define the term "board foot." Discuss how board footage is determined.
2. Group students into pairs. Using several pieces of lumber with different measurements (and labeled with A, B, C, etc.), have the groups determine the board footage of each piece.

METRIC LUMBER MEASURE

1. Review the table in Figure 1-21, page 27, regarding metric lumber sizes. Refer to the Reference Section in the text for additional inch to millimeter conversions used in carpentry.

PANEL MATERIALS

1. Display a variety of panel materials, including plywood (both hardwood and softwood), hardboard, particle board, waferboard, and oriented strand board. Have students determine the composition and construction of each.
2. Have students identify the parts of plywood, including the plies, crossbands, faces, and core. Discuss the difference between exterior and interior plywood.
3. Identify the common thicknesses of plywood and the possible uses of each thickness.
4. Using Figure 1-24, page 29, discuss the groups of

softwood plywood panels that are available. Have the students list the species that fit into each group.

5. Discuss the various plywood veneer grades. Ask students to list applications for the various veneer grades. Once again, stress the importance of using good-quality materials in construction, but not to use material that is too high of a grade.

6. Identify the components of a typical grade-trademark for plywood. Using hypothetical information, have the students place the information in the correct position in the grade-trademark that you have drawn on the chalkboard.

7. Discuss the term "exposure durability classification." Describe how this classification is important in determining the correct type of plywood for a construction project.

8. Discuss the differences between softwood plywood and hardwood plywood grading. Describe how hardwood plywood grades are determined.

9. Using specimens of softwood and hardwood plywood, compare and contrast the construction of each type.

COMPOSITE BOARD

1. Discuss the differences between plywood and composite board. Have students cite different types of composite board.

2. Differentiate between the construction of hardboard, particle board, waferboard, and oriented strand board. Have students cite applications of each type.

3. Obtain samples of different types of composite board. Have students identify each type.

WOOD TREATMENTS

1. Identify different types of wood treatments that are on the market. Determine whether the wood treatments can be classified as oils (such as creosote) or salts that are dissolved in water.

HANDLING AND STORING

1. Stress the importance of proper handling and storage of building materials. Discuss the steps that should be taken for the storage of lumber, panel materials, and interior and exterior finish materials.

2. Discuss precautions for handling of treated lumber.

ENGINEERED WOODS

1. If possible, display samples of engineered lumber and discuss its makeup, applications, and advantages.

NONWOOD MATERIALS

1. Identify various types of nonwood materials that are used in building construction including structural steel members, gypsum lath, wallboard and sheathing, shingles, fasteners, flashing, and caulking materials.

2. Cite recent developments of materials in the construction trades, such as metal structural members. Discuss the types of materials that were originally developed for the commercial trades and are now being used in residential construction. If possible, have examples on hand for inspection.

3. Using Reproducible Master 1-3, *Nail Types, Parts A* and *B,* discuss different nail types used in construction. Discuss similarities and differences between types.

4. Obtain a variety of nails and have students identify each type. Have students cite applications of each type of nail.

5. Using Reproducible Master 1-4, *Nail Sizes,* discuss the size system used for nails.

6. Explain the differences between round, oval, and flat head screws. Identify the dimensions that are used to size screws.

7. Have students list various types of adhesives that are used in the carpentry field, including polyvinyl resin emulsion glue, urea-formaldehyde resin glue, contact cement, and casein glue. Identify similarities and differences between types.

8. Discuss methods for applying adhesives and mastics.

UNIT REVIEW

1. Review the unit objectives. Be sure that the students fully understand each objective.

2. Review Important Terms on page 46, and make certain students are clear on each definition.

3. Assign Test Your Knowledge questions, and Outside Assignments on page 47 of the text. Review the answers to test questions in class. Also, have students report on Outside Assignments as appropriate.

4. Assign pages 7-14 of the workbook. Review the answers in class.

EVALUATION

1. Use Unit 1 Quiz for in-class evaluation. Correct the quizzes and return them to the students for review.

ANSWERS TO TEST YOUR KNOWLEDGE
TEXT PAGE 47

1. lignin
2. Cambium.
3. Willow.
4. edge-grain
5. 15%
6. 30%
7. equilibrium
8. 1 1/2
9. Indoors.
10. B and Better.
11. FAS.
12. 128 bd. ft.
13. Span rating is the maximum recommended center-to-center distance in inches between the supports when the long dimension of the panel is at right angles to the supports. Its purpose is a response to the need to conserve lumber without affecting the strength of structural plywood.
14. Laminated veneer lumber and glue laminated beams.
15. .019″ (See Figure 1-40.)

ANSWERS TO WORKBOOK QUESTIONS
PAGES 7-14

1. lignin
2. A. Heartwood.
   B. Sapwood.
   C. Wood rays.
   D. Cambium.
   E. Bark.
   F. Annual rings.
3. A. cambium
   B. annual or annular rings
4. conifers
5. basswood
   willow
   birch
   ash
6. cypress
7. edge-grained
8. 15%
9. C. 1/30
10. B. 8%
11. radio frequency/condenser
12. A. Round knothole through two wide faces.
    B. Sound encased knot through two wide faces.
    C. Sound, star-checked intergrown knot through two wide faces.
    D. Sound, intergrown knot through two faces.
    E. Sound, intergrown knot through two faces.
    F. Sound, intergrown knot through two faces.
13. Shakes.
14. A. Bow.
    B. Crook.
    C. Twist.
    D. Cup.
15. dimension
16. B and better or supreme
17. FAS
18. D. equilibrium
19. A. Nominal width.
    B. Dressed width.
    C. Face width.
20. Either A. 7 1/4″ or C. 7 1/2″.
21. 832
22. composite
23. B. Southern pine.
24. B. 5/16″
25. B. 40 lb.
26. A. Common.
    B. Box.
    C. Casing.
    D. Finish.
27. C. 3 1/2″
28. A. Round.
    B. Oval.
    C. Flat.
29. A. contact cement
30. B. The glue has a high resistance to moisture.
31. A. Panel grade.
    B. Span rating.
    C. Exposure durability classification.
    D. Mill number.
32. Span rating
33. truss joist
34. 40
35. Purpose is to hold parts of a building frame together. They anchor frames to foundations, floors to walls, and roofs to walls. They are especially important in areas having earthquakes or high winds.

ANSWERS TO UNIT 1 QUIZ

1. False.
2. True.
3. False.
4. False.
5. True.
6. True.
7. C. plain-sawed
8. D. Particleboard.
   G. Waferwood.
9. D. softwood

10. C. 3.33
11. C
12. A
13. B
14. D

Reproducible Master 1-1

Basic House Parts

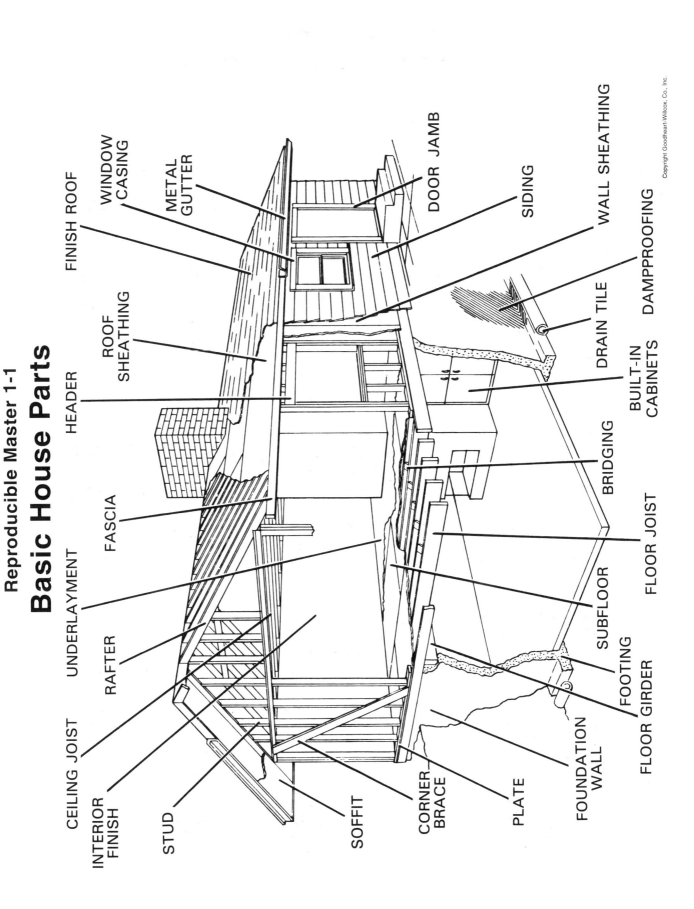

WINDOW CASING

FINISH ROOF

METAL GUTTER

DOOR JAMB

SIDING

WALL SHEATHING

ROOF SHEATHING

HEADER

DAMPPROOFING

DRAIN TILE

FASCIA

BUILT-IN CABINETS

BRIDGING

UNDERLAYMENT

RAFTER

FLOOR JOIST

SUBFLOOR

CEILING JOIST

INTERIOR FINISH

STUD

SOFFIT

CORNER BRACE

PLATE

FOUNDATION WALL

FOOTING

FLOOR GIRDER

29

Reproducible Master 1-2
Typical Grade Stamps

Dimension Grades

12 ⓌWP® **2** S-DRY △D.FIR

12 ⓌWP® **CONST** S-DRY **ES-AF**

12 ⓌWP® **STAND** &BTR S-DRY △D.FIR S

Commons

12 ⓌWP® **2&BTR COM** S-DRY ΦΦ

12 ⓌWP® **4 COM** S-DRY ⬡ES

12 ⓌWP® **STERLING** S-DRY IWP

Glued Products

12 ⓌWP® **STUD** S-DRY STUD USE ONLY CERT GLUED JNTS WW

12 ⓌWP® **1** S-DRY CERT EXT JNTS HEM FIR

Machine Stress-Rated Products

MACHINE RATED
ⓌWP® 12 S-DRY △D FIR
1650 Fb 1.5E

Finish & Select Grades

12 ⓌWP® **C & BTR SEL** MC 15 ΦΦ

12 ⓌWP® **PRIME** MC 15 △D FIR

12 ⓌWP® **D SEL** MC 15 SP

Finish Grade — Graded Under WCLIB Rules

12 ⓌWP® **C & BTR** VG S-DRY WCLB RULES △D FIR

Cedar Grades

12 ⓌWP® **CLEAR** VG **HEART** MC15 WR CDR

12 ⓌWP® **A** MC15 WESTERN CEDAR

Decking

12 ⓌWP® **SEL DECK** MC 15 INC CDR

Species Identification

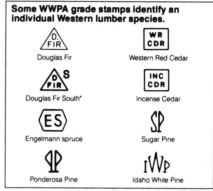

Some WWPA grade stamps identify an individual Western lumber species.

△D.FIR — Douglas Fir
WR CDR — Western Red Cedar

△D.FIR S — Douglas Fir South*
INC CDR — Incense Cedar

ⒺES — Engelmann spruce
SP — Sugar Pine

ΦΦ — Ponderosa Pine
IWP — Idaho White Pine

*Lumber manufactured from Douglas Fir grown in Arizona, Colorado, New Mexico and Utah.

A number of Western lumber species have similar performance properties and are marketed with a common species designation.

DOUG. FIR-L — Douglas Fir and Larch

HEM FIR — California Red Fir, Grand Fir, Noble Fir, Pacific Silver Fir, White Fir and Western Hemlock

ΦPSP — Ponderosa Pine and Sugar Pine

ES-AF — Engelmann Spruce and Alpine Fir

WESTERN CEDAR — Incense and Western Red Cedar

Because of timber stand composition, some mills market additional species combinations.

ES LP — Engelmann Spruce, Lodgepole Pine

WESTERN WOODS — Western Woods (Any combination of western softwood species except Redwood)

WW — White Woods (Engelmann Spruce, and true firs, any hemlocks, and pines)

PP-LP — Pondersoa Pine, Lodgepole Pine

ES LP AF — Engelmann Spruce-Alpine Fir-Lodgepole Pine

Western Wood Products Association

Reproducible Master 1-3 (Part A)
Nail Types

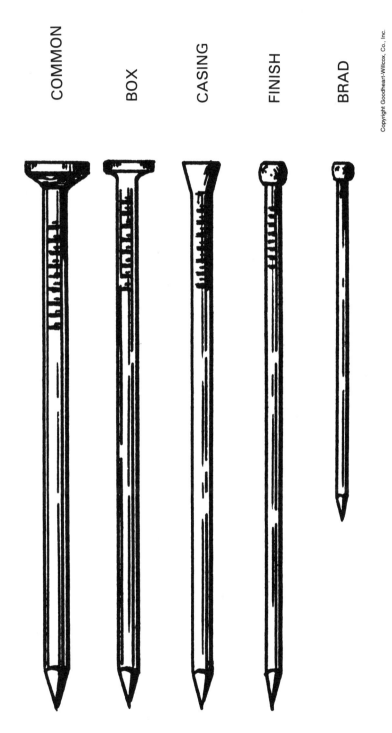

COMMON

BOX

CASING

FINISH

BRAD

Reproducible Master 1-3 (Part B)
Nail Types

NAIL FOR GENERAL USE

NAIL FOR GENERAL USE

TRUSSED RAFTER NAIL

POLE-CONSTRUCTION NAIL

FLOORING NAIL

UNDERLAY FLOOR NAIL

DRYWALL NAIL

ROOFING NAIL WITH
NEOPRENE WASHER

ROOFING NAIL WITH
NEOPRENE WASHER

ASPHALT SHINGLE NAIL

ASPHALT SHINGLE NAIL

WOOD SHINGLE FACE NAIL

ENAMELED FACE NAIL FOR
INSULATED SIDING, SHAKES

NAIL FOR APPLYING
SIDING TO PLYWOOD

NAIL FOR APPLYING
ROOFING TO PLYWOOD

DUPLEX-HEAD NAIL

Reproducible Master 1-4
Nail Sizes

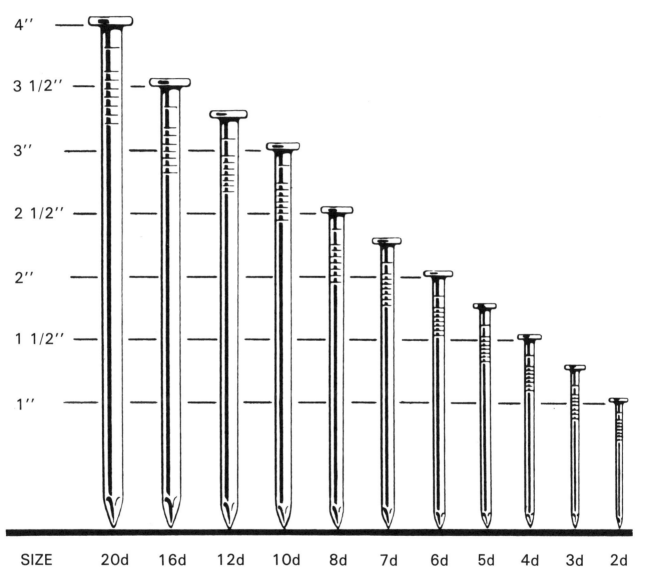

SIZE   20d   16d   12d   10d   8d   7d   6d   5d   4d   3d   2d

Unit 1 Quiz
Building Materials

Name _____ Score _____

True-False

Circle T if the answer is True or F if the answer is False.

T   F   1. Knots are separations along the grain and between the annular growth rings.

T   F   2. FAS is the best grade of hardwood lumber.

T   F   3. Plywood always contains an even number of layers.

T   F   4. Nail size is measured in a unit called "penny," which is abbreviated with the letter "p."

T   F   5. Most metal structural materials are color-coded so that different gauges are not mixed at the construction site.

T   F   6. Hardboard is made of refined wood fibers, pressed together to form a hard, dense material.

Multiple Choice

Choose the answer that correctly completes the statement. Write the corresponding letter in the space provided.

_____   7. Hardwood lumber that is cut so the annular rings form an angle less than 45° with the surface is called _____ lumber.
   A. edge-grained
   B. quarter-sawed
   C. plain-sawed
   D. flat-grained

_____   8. Which of the following is/are *not* included in the group known as "engineered lumber?"
   A. Laminated veneer lumber.
   B. Wood I-beams.
   C. Glue-lams.
   D. Particleboard.
   E. Open-web truss.
   F. Parallel strand lumber.
   G. Waferwood.
   H. Laminated strand lumber.

_____   9. Basic classifications of _____ grading includes boards, dimension, and timbers.
   A. hardwood plywood
   B. softwood plywood
   C. hardwood
   D. softwood

_____ 10. A piece of stock measuring 1″ × 10″ × 4′ contains _____ bd. ft.
       A. .27
       B. 2.7
       C. 3.33
       D. 4

Identification

Identify the following types of warp.

_____ 11. Twist.

_____ 12. Bow.

_____ 13. Crook.

_____ 14. Cup.

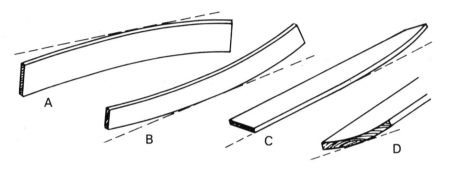

2

General Safety Rules

OBJECTIVES

Students will be able to:

- ❑ List and describe appropriate clothing suitable for carpenters to wear in safety.
- ❑ List other personal safety equipment recommended for carpenters to use.
- ❑ Cite safety measures relating to hand and power tools.
- ❑ Explain housekeeping measures designed to promote safe working conditions.
- ❑ List safety measures relating to shoring and scaffolding.
- ❑ Detail safe practices in handling and working pressure-treated lumber.
- ❑ Describe proper methods of lifting and carrying that avoids personal injury.
- ❑ Describe classes of fires.
- ❑ Recommend practices that help prevent fires.

INSTRUCTIONAL MATERIALS

Text: Pages 51-54
   Important Terms, page 54
   Test Your Knowledge, page 54
Workbook: Pages 15-16
Instructor's Manual:
   Reproducible Master 2-1, *Safety: Fall Protection*
   Unit 2 Quiz

TRADE-RELATED MATH

It is a fact that objects in a free fall accelerate at the rate of 32 feet per second for each second elapsed from their resting height to the ground. For example, if a hammer dropped from the roof of a two-story-building takes 6 seconds to reach the ground, how fast will it be falling at the point of impact? The answer may be found by adding:

$$32 + 32 + 32 + 32 + 32 + 32 = 192$$

Or, by multiplying $32 \times 6 = 192$

INSTRUCTIONAL CONCEPTS AND STUDENT LEARNING EXPERIENCES

CLOTHING

1. Discuss the reasons for special clothing requirements and possible consequences for ignoring these rules.
2. Have students evaluate their fellow students' clothing with respect to safety rules.
3. Demonstrate a hard hat's effectiveness in preventing injury from falling objects by dropping a tool on it from different heights.

PERSONAL PROTECTIVE EQUIPMENT

1. Using a pair of approved safety glasses, discuss their properties and effectiveness. Suggest the types of flying objects that carpenters might encounter while working.
2. Discuss the effectiveness of safety shoes with reinforced toes in preventing crushing injuries.
3. Display a pair of gloves that could be worn to avoid injuries from rough materials.

HAND AND POWER TOOL SAFETY

1. Set up a demonstration to show the merits of using sharp tools.
2. Demonstrate the proper way to use various hand tools common to the carpenter's trade.
3. Demonstrate the safe and proper method of operating various power tools.
4. Refer students to the safety sections of Unit 4. Suggest that they study these rules in preparation for later use of the power tools.

GOOD HOUSEKEEPING

1. Review the rules of good housekeeping on the job site, pointing out the dangerous consequences represented by stumbling or slipping over debris.
2. If possible, visit a construction site; have students

note any housekeeping problems. Ask students to write down their evaluation of the job site and how problems might be remedied.

EXCAVATIONS

1. Explain the dangers of working in excavations not properly sloped or shored up.
2. If possible, show photos of excavations properly sloped or shored.

SCAFFOLDS AND LADDERS

1. Explain the factor of four to students and show pictures of appropriate, approved scaffolds.
2. Using Reproducible Master 2-1, *Safety: Fall Protection,* discuss the OSHA regulations regarding avoidance of falls and safeguards required on scaffolds.
3. When on a job site, make sure students check the safety of any scaffolds and ladders used.
4. Refer to Unit 29, *Scaffolds and Ladders*, and have students review safety information contained there.

FALLING OBJECTS

1. Stress the need for good housekeeping on scaffolds and on upper levels of buildings.
2. Demonstrate the force of a falling object and suggest the injury that could result from being struck.
3. Stress the need for vigilance in walking underneath carpenters at work on an upper level.

HANDLING PRESSURE-TREATED LUMBER

1. Discuss the toxic properties of water borne and oil-borne preservatives used on lumber.
2. Demonstrate the proper safeguards to use in working with pressure-treated lumber.

LIFTING AND CARRYING

1. Demonstrate the proper method for lifting and carrying heavy loads.
2. Suggest that assistance should be sought for lifting and carrying very heavy objects.
3. Have students practice proper lifting and carrying techniques.

FIRE PROTECTION

1. Review the types of fires. Relate this review to the proper types of extinguishers that should be used for each type of fire.
2. Explain the relationship of fire prevention to good housekeeping.
3. Show examples of approved containers for storing of combustible materials.

4. Make certain that students know the location of fire extinguishers and that they know how to use them.

FIRST AID

1. Maintain a fully equipped first aid kit and drill students on basic first aid procedures.
2. Keep the kit available at all times and make certain students know its location.

UNIT REVIEW

1. Review the unit objectives. Be sure students fully understand each one.
2. Assign *Important Terms* and *Test Your Knowledge* questions on page 54 of the text. Review answers in class.
3. Assign pages 15-16 of the workbook. Review the answers in class.

EVALUATION

1. Use Unit 2 Quiz for in-class evaluation. Correct the quizzes and return them to the students for review.

ANSWERS TO TEST YOUR KNOWLEDGE TEXT PAGE 54

1. B. Trousers or overalls without cuffs.
2. Whenever work involves the slightest hazard to the eyes.
3. C. Wherever there is a danger from falling objects.
4. C. 2500 lb.
5. D. Stored in panels or chests.
6. False.

ANSWERS TO WORKBOOK QUESTIONS PAGES 15-16

1. carefully, safety rules
2. Leather soled
3. C. A 1/8" diameter ball dropped from 50"
4. B. 80 lb. ball dropped 5'
5. True.
6. B. day
7. C. class C
8. additional force
9. neatness
10. B. special

ANSWERS TO UNIT 2 QUIZ

1. True.
2. False.

3. True.
4. False.
5. False.
6. All of the above.
7. D. wear a dust mask and do sawing outdoors
8. A. turn or twist
9. volatile
10. housekeeping

Reproducible Master 2-1
Safety: Fall Protection

Name _____ Period _____

OSHA's new fall protection regulations cover seven different job-site areas.

Floor Framing	Do not stand on sill or top plates while placing joists or floor trusses. Work from the ground, ladders, or sawhorse staging.
Subfloor	Install the first course from the ground, ladders, or staging, then work from the deck. Paint a warning line six feet from the edges that are unprotected.
Working on Decks	Workers not doing leading-edge work (including those cutting material) must stay six feet in from the edges or other fall hazards.
Outside Walls	Complete as much work as possible away from the edge of the deck. Try to store materials within the six foot warning perimeter.
Trussing and Rafter Framing	Set first two trusses or rafters from ladders or staging. Brace trusses and rafters before using them for support. If they can not safely work from staging, workers can then work from the top plate, using stabilized roof framing for support.
Roof Sheathing	Install bottom course of sheathing while standing in the truss webs. Then install slide guards, at least four inches high, across the roof at intervals of thirteen feet. Steeper roofs will require higher guards and closer intervals.
Guard Rails and Openings	Block floor and wall openings with guard rails. Set top rail of guard forty-two inches off the subfloor and a midrail at around twenty-one inches. Install a toe guard on the deck. Window openings lower than thirty-nine inches also require a rail.

Unit 2 Quiz
General Safety Rules

Name _____ Score _____

True-False

Circle T if the answer is True or F if the answer is False.

T   F   1. Working safely depends on the willingness to follow the safety rules.

T   F   2. It is permissible and safe to wear lightweight canvas shoes when doing carpentry work in a very warm climate.

T   F   3. Class C fires are caused by electrical wiring and equipment.

T   F   4. Upon arriving on the job site, it is good practice to set out all of your tools so they will be close at hand. This makes the carpenter safer and more productive.

T   F   5. It is not advisable to keep a first aid kit on the job site because of dirt, dust, and generally unsanitary conditions.

Multiple Choice

Choose the answer that correctly completes the statement. Write the corresponding letter in the space provided.

_____ 6. Under what conditions is shoring required for excavating?

A. When the hole is more than 5′ deep.
B. If a caving is likely to cover workers.
C. When the ground is cracked or caving is likely to occur.
D. All of the above.

_____ 7. When sawing pressure treated lumber, _____.

A. avoid prolonged unprotected breathing where there is airborne sawdust
B. thoroughly wash any skin that has come in contact with the chemicals
C. wear a dust mask and saw outdoors
D. All of the above.

_____ 8. When carrying a heavy load, do not _____ your body but make adjustments in position by shifting your feet.

A. turn or twist
B. bend
C. straighten
D. None of the above.

Completion

Write the answer that correctly completes the statement in the space provided.

_____ 9. Always keep containers of _____ materials closed when not in use.

_____ 10. Good _____ refers to the neatness and good order of the construction site.

3

Hand Tools

OBJECTIVES

Students will be able to:

❏ Recognize the most common hand tools.
❏ Select the proper hand tool for a given job.
❏ Identify the main parts of each major hand tool.
❏ Explain appropriate methods of tool maintenance and storage.

INSTRUCTIONAL MATERIALS

Text: Pages 55-74
  Important Terms page 72
  Test Your Knowledge, page 72
  Outside Assignments, page 73
Workbook: Pages 17-22
Instructor's Manual:
  Reproducible Master 3-1, *Saw Tooth Shapes*
  Reproducible Master 3-2, *Parts of a Plane*
  Reproducible Master 3-3, *Parts of a Claw
      Hammer*
  Unit 3 Quiz

TRADE-RELATED MATH

Working with fractions is required in almost every aspect of carpentry, from rough layout to finish work. The most common fractional increment is the sixteenth of an inch. In order to work math computations with fractions, a common denominator must be first obtained. A common denominator is determined by multiplying the upper (numerator) and lower (denominator) part of the fraction by the same number. For example:

$$\frac{1}{2} = \frac{x}{8} \qquad \frac{3}{8} = \frac{x}{16}$$

$$\frac{1}{2} \times \frac{4}{4} = \frac{4}{8} \qquad \frac{3}{8} \times \frac{2}{2} = \frac{6}{16}$$

Once a common denominator has been determined for the fractions in the problem,

computation may begin. When adding fractions, the denominator remains the same, while the numerator is added. For example:

$$\frac{1}{2} + \frac{1}{4} = x$$

$$\frac{1}{2} \times \frac{2}{2} = \frac{2}{4}$$

$$\frac{2}{4} + \frac{1}{4} = \frac{3}{4}$$

When subtracting fractions, a common denominator must first be determined. The denominator remains the same, while the numerators are subtracted. For example:

$$\frac{9}{16} - \frac{1}{4} = x$$

$$\frac{1}{4} \times \frac{4}{4} = \frac{4}{16}$$

$$\frac{9}{16} - \frac{4}{16} = \frac{5}{16}$$

INSTRUCTIONAL CONCEPTS AND STUDENT LEARNING EXPERIENCES

MEASURING AND LAYOUT TOOLS

1. Review the layout tools commonly used by a carpenter. Compare and contrast these to tools that the students may have used in a woodworking or cabinetmaking course. Explain why it is important for a carpenter to use good-quality tools.

2. If necessary, review the markings on a tape rule or a folding rule. Compare a conventional inch-foot tape with an SI metric tape. Explain the division marks as necessary.

3. Demonstrate the use of layout tools that students are not accustomed to using such as the framing square, marking gauge, level, and plumb line. Ask the students to show the proper use of the tools after the demonstration.

SAWS

1. Identify the hand saws commonly used by a carpenter. Review the applications of each by demonstrating their use.
2. Using Reproducible Master 3-1, *Saw Tooth Shapes,* compare and contrast the saw tooth shapes of the crosscut saw (A) and ripsaw (B). Have the students explain why different types of teeth are necessary on crosscut and ripsaws.
3. Have students list the types of saws commonly used for rough cutting and finish work. Ask the students to explain their categorization.
4. Discuss the different types of set used for hacksaw blades.
5. Stress the importance of using saws safely.

PLANING, SMOOTHING, AND SHAPING

1. Review the types of planing and smoothing tools used by a carpenter.
2. Identify the parts of the tools (use Reproducible Master 3-2, *Parts of a Plane.*)
3. Emphasize the importance of using proper terminology when discussing tools.
4. Demonstrate the proper use of planing and smoothing tools with which students are not familiar. Stress the importance of using the tools safely.

DRILLING AND BORING

1. Review the types of drilling and boring tools commonly used by a carpenter. Discuss the types and sizes of bits that can be used with the brace, push drill, and hand drill.
2. Explain how drills are sized and purchased in a store.
3. Stress the importance of using the drilling and boring tools safely.

FASTENING PARTS TOGETHER

1. Review the types of tools that carpenters commonly use to fasten parts together.
2. Distinguish between the curved claw and rip claw hammers. Ask the students if there are any applications where one type of hammer should be used.
3. Review the parts of a hammer using Reproducible Master 3-3, *Parts of a Hammer,* and discuss the proper procedure for driving and pulling nails with the hammer.
4. Have the students check the condition of the hammers in the shop. With instructor supervision, ask the students to repair any tools that are in disrepair.

5. Identify the tools that students may not be familiar with, such as hatchets, ripping bars, tackers, staplers, and nailers. Discuss applications of these tools, and demonstrate the proper, safe use of these tools.
6. Discuss the criteria that should be used to select an appropriate screwdriver for the given application (width and thickness of the blade).

CLAMPING TOOLS

1. Identify the two types of clamping tools that are commonly used by a carpenter—C-clamps and hand screws. Ask students to demonstrate the proper use of these tools.

TOOL STORAGE

1. Discuss the importance of having some type of chest or cabinet to store tools.
2. Have the students design some type of storage chest for their tools. Have the students evaluate each other's designs based on organization of the tools and preventing damage to their cutting edges.

CARE AND MAINTENANCE

1. Discuss the means of properly maintaining tools.
2. Demonstrate the proper method used to hone edge tools on an oilstone.
3. Have the students identify tools in the shop that may need maintenance. With instructor supervision, have students repair or maintain these tools.

GENERAL SAFETY RULES

1. Discuss the General Safety Rules in Unit 2 of the text. Once again, stress the importance of working safely in class and on the job site. Using statistics obtained from the Occupational Safety and Health Administration (OSHA), emphasize the fact that a person working safely with less productivity is of more value to a company than a worker who is injured on the job due to unsafe work habits.

UNIT REVIEW

1. Review the unit objectives. Be sure that the students fully understand each objective.
2. Have the students develop their own lists of hand tools that they feel should be included in a carpenter's toolbox. Make sure that they understand the importance of having enough tools, but not every tool that is available. Have the students explain their tool selections.

3. Assign *Important Terms, Test Your Knowledge* questions, and *Outside Assignments* on pages 72 and 73 of the text. Review the answers in class.
4. Assign pages 17-22 of the workbook. Review the answers in class.

EVALUATION

1. Use Unit 3 Quiz for in-class evaluation. Correct the quizzes and return them to the students for review.
2. Given a set of tools that need to be repaired, have the students identify the problems with the tools and properly repair them.

ANSWERS TO TEST YOUR KNOWLEDGE
TEXT PAGE 72

1. 6
2. C. 16″
3. Level.
4. False.
5. D. 14-16 teeth per inch
6. up
7. 5/8
8. False.
9. D. weight of the head
10. Curved claw—claw used to pull nails. Ripping (straight claw)—claw used to pry apart fastened pieces.
11. To cut light sheet metal and asphalt shingles.
12. Slotted—by length of the blade.
    Phillips—as a point number ranging from No. 0, the smallest, to No. 4, the largest.
13. inside
14. jointing

ANSWERS TO WORKBOOK QUESTIONS
PAGES 17-22

1. A. tape rule
   B. folding wood rule
2. quick square
   super square
3. A. Backsaw.
   B. Crosscut saw.
   C. Drywall saw.

4. Seven.
5. D. 14″
6. block plane
7. miter box
8. B. 3/8
9. brace
10. expansive
11. push drill
12. rip claw
13. A. Head.
    B. Face.
    C. Neck.
    D. Cheek.
    E. Claw.
    F. Handle.
14. shingling hatchets
15. B. thirty-seconds
16. A. ferrule
17. A. Phillips.
    B. Standard.
    C. Cabinet.
18. A. hand screw
19. A. 30°-35°
    B. 25°-30°
20. inside
21. D. Jointing
22. B. spiral ratchet screwdriver
23. B. OSHA
24. can

ANSWERS TO UNIT 3 QUIZ

1. False.
2. True.
3. True.
4. False.
5. True.
6. True.
7. False.
8. False.
9. True.
10. False.
11. C. rafter square
12. B. T-bevel
13. D. face
14. B. rounded

Saw Tooth Shapes

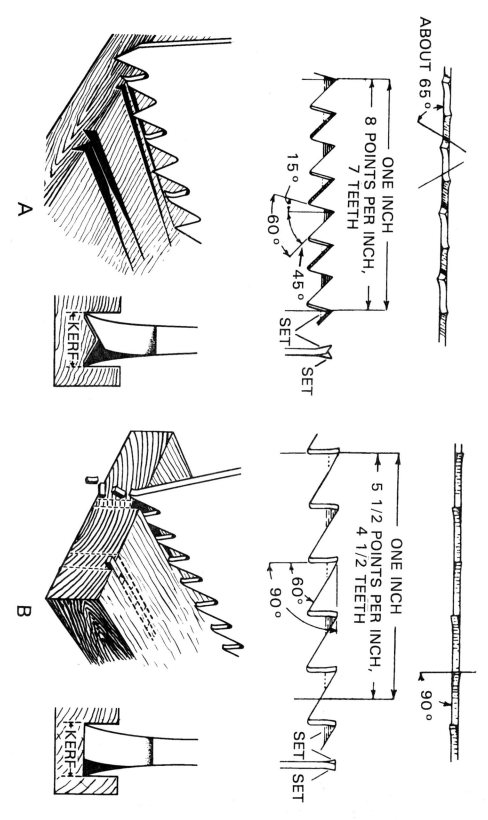

ABOUT 65°

ONE INCH
8 POINTS PER INCH,
7 TEETH

15°
60°
45°

SET
SET

A

KERF

ONE INCH
5 1/2 POINTS PER INCH,
4 1/2 TEETH

60°
90°

90°

SET
SET

B

KERF

Reproducible Master 3-1
Saw Tooth Shapes

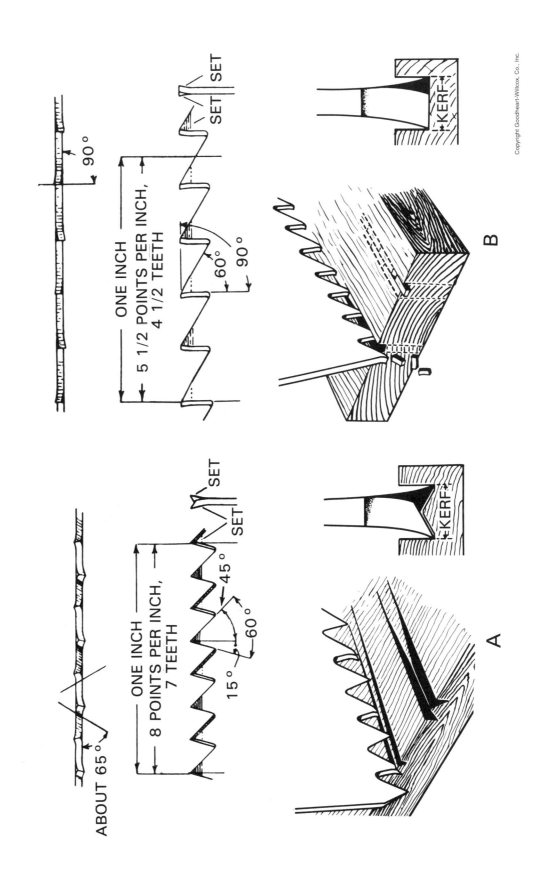

ABOUT 65°

ONE INCH

8 POINTS PER INCH, 7 TEETH

45°
60°
15°

SET

SET

KERF

A

90°

ONE INCH

5 1/2 POINTS PER INCH, 4 1/2 TEETH

60°
90°

SET | SET

KERF

B

Reproducible Master 3-2
Parts of a Plane

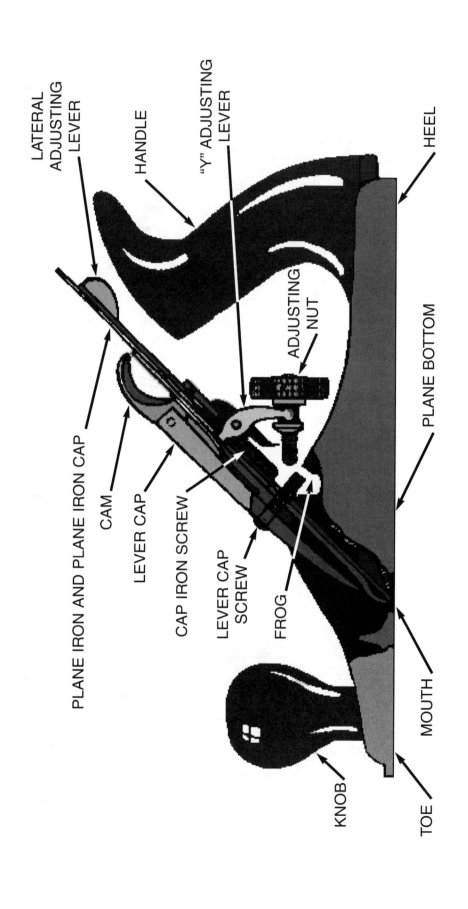

LATERAL ADJUSTING LEVER

HANDLE

"Y" ADJUSTING LEVER

HEEL

ADJUSTING NUT

PLANE BOTTOM

PLANE IRON AND PLANE IRON CAP

CAM

LEVER CAP

CAP IRON SCREW

LEVER CAP SCREW

FROG

KNOB

MOUTH

TOE

Parts of a Claw Hammer

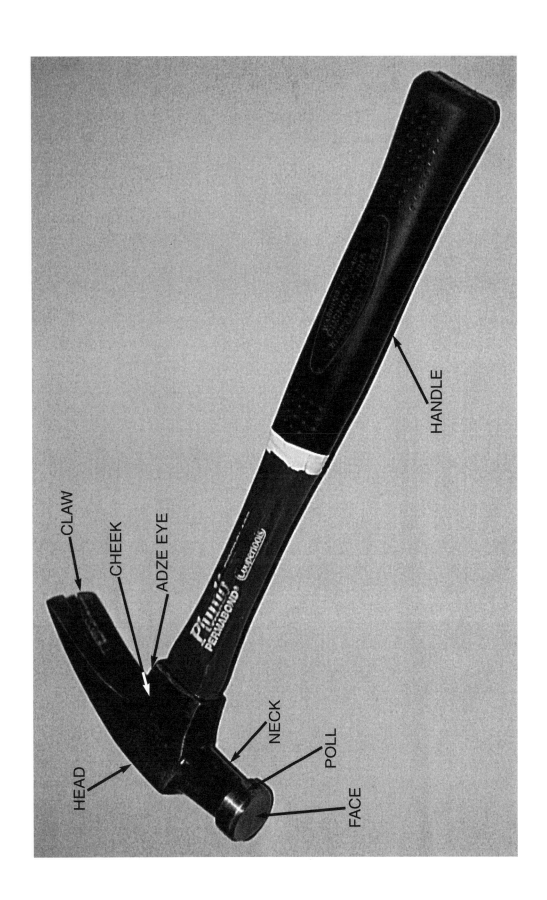

CLAW

CHEEK

ADZE EYE

HANDLE

NECK

POLL

FACE

HEAD

Unit 3 Quiz
Hand Tools

Name _____ Score _____

True-False
Circle T if the answer is True or F if the answer is False.

T   F   1. A saw used for finishing work should have fewer teeth per inch than one used for general-purpose applications.

T   F   2. A plumb bob is used to establish a vertical line.

T   F   3. The size of a hand drill is determined by the capacity of its chuck.

T   F   4. A nail set is used to drill holes for casing or finishing nails.

T   F   5. Sizes of Phillips screwdrivers range from No. 0 to No. 4, with No. 4 being the largest.

T   F   6. The ripping claw hammer has a straight claw.

T   F   7. The blade of a block plane is set at a higher angle than a jack plane to produce a finer cut.

T   F   8. Expansive bits are fixed-size bits.

T   F   9. Hacksaw blades should be installed so the teeth point away from the handle.

T   F 10. A super square is shaped like a conventional framing square, but is larger.

Multiple Choice
Choose the answer that correctly completes the statement. Write the corresponding letter in the space provided.

_____ 11. The framing square is also called a _____.

    A. combination square
    B. try square
    C. rafter square
    D. T-bevel

_____ 12. The _____ has/have an adjustable blade and is used to transfer angles from one place to another.

    A. try square
    B. T-bevel
    C. wing dividers
    D. combination square

_____ 13. The part of a hammer that contacts a nail when driving it is the _____.

    A. adze eye
    B. cheek
    C. neck
    D. face

_____ 14. Gullets of a drywall saw are _____ to prevent their clogging from the gypsum material.

    A. squared off
    B. rounded
    C. tapered
    D. All of the above.

4

Power Tools

OBJECTIVES

Students will be able to:
- ❏ Recognize common power tools.
- ❏ Explain the function and operation of the principle power tools.
- ❏ Identify the parts of common power tools.
- ❏ Apply safety rules.

INSTRUCTIONAL MATERIALS

Text: Pages 75-100
    Important Terms, page 99
    Test Your Knowledge, page 99
    Outside Assignments, page 99
Workbook: Pages 23-38
Instructor's Manual:
    Reproducible Master 4-1, *Portable Power*
        Circular Saw
    Reproducible Master 4-2, *Jointer Operation*
    Unit 4 Quiz

TRADE-RELATED MATH

In some instances, a carpenter must be able to use decimals. Measurements on a plot plan, for example, may be shown in a decimal format. In order to add or subtract decimal values, always make sure that the decimal points align, no matter how many numbers are shown on each side of the decimal points. (Mention that as many zeros as desired can be added after the last number behind the decimal point, without altering its value.) To add decimal values, simply align the decimal points and add as shown in the following example:

$$400.26 + 25.7 = x$$

$$\begin{array}{r} 400.26 \\ +25.70 \\ \hline 425.96 \end{array}$$

To subtract fractions, once again align the decimal points, but this time subtract the values.

$$345.76 - 221.52 = x$$

$$\begin{array}{r} 345.76 \\ -221.52 \\ \hline 124.24 \end{array}$$

$$x = 124.24$$

In some cases, numbers must be carried over (for addition) or borrowed from (subtraction) in order to add or subtract.

INSTRUCTIONAL CONCEPTS AND STUDENT LEARNING EXPERIENCES

POWER TOOL SAFETY

1. Review the General Safety Rules for the use of power tools in Unit 2 of the text. Emphasize the importance of working safely in class and on the job site.
2. Stress the fact that many of the tasks performed by a carpenter require the use of power tools. Instruct students to think through an operation before performing it. Demand that students have their total focus placed on the operation they are about to perform. Discourage students from distracting others during an operation.

ELECTRICAL SAFETY

1. Describe the function of magnetic starters on power tools. Stress that power tools should be in the "off" position while in storage and before they are plugged into a live circuit.
2. Emphasize the importance of using grounded "three-wire" cords and plugs with power tools.
3. Identify potential electrical problems that may be evident in a power tool including frayed cords, grounding plugs that have been broken off, etc.

Have students identify possible "fixes" for these problems. Stress that a *cheap* repair is not necessarily the best type of repair.

4. Define the term "ground fault circuit interrupter" as it relates to electrical safety.

5. Have the students inspect the electrical cords of the power equipment. With instructor supervision, have the students make any necessary repairs.

PORTABLE CIRCULAR SAWS

1. Using Reproducible Master 4-1, *Portable Power Circular Saw*, list the types of portable circular saws and then identify the parts. Have the students note the location and function of each part on both types of saws.

2. Discuss the different types of blades that can be used on a portable circular saw, citing the applications of each. Stress the importance of selecting the proper blade for the task being performed.

3. Review the safety rules for using and servicing portable circular saws on pages 77-80 of the text.

4. Demonstrate the proper use of a portable circular saw. Use the saw to cut different types of materials, making straight and angled cuts.

SABER SAWS

1. Identify the parts of a saber saw. Have the students cite applications where a saber saw might be used.

2. Show the students a variety of saber saw blades. Have them describe possible applications of each.

3. Review the safety rules for using and servicing saber saws on pages 81-82 of the text.

4. Demonstrate the proper, safe use of a saber saw. Have students note where splintering occurs (on the topside of the work). Show the students the proper methods of cutting internal openings.

PORTABLE ELECTRIC DRILLS

1. Identify the parts of the portable electric drill. Discuss how rpm speeds vary and how chuck capacity determines the size of a drill.

2. Identify the different types of drill bits that may be used with electric and cordless drills.

3. Review the safety rules for using and servicing portable electric drills on pages 82-84 of the text.

4. Demonstrate the safe use of a portable electric drill. Stress that the pressure required to operate a drill will vary with the drill size, type of wood, and the diameter of the drill bit.

POWER PLANES

1. Identify the parts of a power plane. Compare the parts of a power plane with those of a hand plane.

2. Review the safety rules for using and servicing power planes on page 85 of the text.

3. Demonstrate the safe use of a power plane. Show the students how to plane chamfers and make square cuts.

PORTABLE ROUTERS

1. Identify the parts of a portable router.

2. Review the safety rules for using portable routers on page 88 of the text.

3. Demonstrate the safe use of a router. Stress the importance of moving the router from left to right along an edge when making a cut along an edge. Show the students how to use bits with and without pilot tips.

PORTABLE SANDERS

1. Identify the three basic types of portable sanders: belt, disc, and finish. Discuss the advantages and disadvantages of each.

2. Name the parts of portable sanders and have students identify the location of each.

3. Differentiate between orbital and oscillating finish sanders.

4. Demonstrate the safe use of portable sanders.

STAPLERS AND NAILERS

1. Cite advantages of using power staplers and nailers over using hand tools to fasten materials. Stress the safety concerns that must be recognized.

2. Review the safety rules for power staplers and nailers on page 89 of the text.

3. Demonstrate the proper use of power nailers and staplers. Stress the importance of making sure that the nail or staple "hits its mark" in the material, thus, eliminating the possibility of stray nails or staples.

RADIAL ARM SAWS

1. Identify the parts of the radial arm saw. Cite applications where a radial arm saw might be used.

2. Review the safety rules for using radial arm saws on page 92 of the text. Emphasize that the saw may tend to *"feed itself"* into the stock, and that the saw handle should be grasped firmly to control the saw.

3. Demonstrate the safe use of a radial arm saw. In addition to demonstrating crosscutting and ripping operations, show the students how to cut miters and bevels.

TABLE SAWS AND JOINTERS

1. Review the parts of a table saw. Have students identify operations that can be performed on the table saw other than crosscutting and ripping operations.
2. Discuss preparations that must be made to the stock before cutting it on the table saw.
3. Review the safety rules for using and servicing the table saw on pages 93-95 of the text. Stress the importance of using the guard for most cutting operations.
4. Have students design their own "personalized" push sticks. Emphasize the margin of safety that should be maintained between the saw operator's hands and the saw blade.
5. Identify the parts of the jointer. Compare the parts and operation of a jointer to those of a power plane.
6. Using Reproducible Master 4-2, *Jointer Operations*, review the safety rules for using a jointer on pages 95- 96 of the text. Emphasize the minimum dimensions of stock that can be safely jointed.
7. Demonstrate the safe use of the jointer. Point out the importance of "stepping" hands along the stock to avoid bearing down on the stock while it passes over the cutter head.

SPECIAL SAWS

1. Identify special saws that are commonly used by a carpenter. Include the power miter saw, the plate joiner, the frame and trim saw, and the chain saw. Discuss why these saws are now often found in the carpenter's tool chest.
2. Identify the principle parts of the power miter box and the frame and trim saw. Discuss the function of each part.
3. Review the safety rules for using specialty saws on page 97 of the text. In addition, refer to the operator's manuals for specific operating and safety instructions.

SPECIALTY TOOLS

1. Identify tools designed for special carpentry operations. Include tools such as drywall screw guns, plate joiners, panel saws, and powder-actuated tools.
2. If possible, demonstrate the use of several specialty tools.

POWER TOOL CARE AND MAINTENANCE

1. Emphasize the importance of properly maintaining power tools.
2. Have the students identify power tools in the shop that need repairing. With instructor supervision, have students make any necessary repairs.

UNIT REVIEW

1. Review the unit objectives. Be sure that the students fully understand each objective.
2. Have the students develop lists of power tools that they feel should be purchased by a carpenter. Emphasize that having the necessary tools does *not* require that they own *every* tool available. Have the students explain their tool selections.
3. Assign *Important Terms, Test Your Knowledge* questions, and *Outside Assignments* on page 99 of the text. Review the answers and reports in class.
5. Assign pages 23-28 of the workbook. Review the answers in class.

EVALUATION

1. Use Unit 4 Quiz for in-class evaluation. Correct the quizzes and return them to the students for review.

ANSWERS TO TEST YOUR KNOWLEDGE TEXT PAGE 99

1. Device which turns off electrical power if a wire in a tool grounds out (shorts) against the tool's case. The device senses the ground and trips at 5 milliamperes.
2. Check the label on the tool's case.
3. Ground continuity involves the third or ground conductor which will bleed off current resulting from a short in a tool's housing or case. It protects the operator from electrical shock.
4. blade diameter
5. 10
6. up
7. False.
8. front shoe
9. chain saw
10. collet
11. belt width
12. overhead arm
13. It's faster and it allows working in close quarters.
14. False.
15. pulled toward

16. outfeed table
17. All except rip cuts.

ANSWERS TO WORKBOOK QUESTIONS
PAGES 23-28

1. stationary
2. grounded
3. A. Portable circular saw.
   B. Belt sander.
   C. Router.
   D. Saber saw.
   E. Finishing sander.
4. off
5. A. Switch.
   B. Motor.
   C. Shoe or base.
   D. Telescoping Guard.
   E. Arbor and Locking Bolt.
   F. Upper Guard.
6. shoe
7. A. 1/8″
8. A. Rough cut combination.
   B. Rip.
   C. Crosscut.
   D. Standard combination/miter.
9. upward
10. B. 1/2″
11. C. capacity of chuck
12. B. Place base of drill firmly on stock before starting motor.
13. D. 20,000
14. A. front shoe
15. collet
16. clockwise
17. B. belt width
18. B. compressed air
19. C. overhead arm
20. A. Saw feed.
    B. Stock.
    C. Table.

D. Thrust.
E. Fence.
21. C. 6″
22. toward
23. A. Guard and splitter.
    B. Miter gauge.
    C. Table.
    D. Fence.
    E. Saw raising handwheel.
    F. Saw tilt handwheel.
24. B. When ripping stock free hand, do not use the fence.
25. A. Rear outfeed table.
    B. Fence.
    C. Front infeed table.
    D. Depth scale.
    E. Front guard.
    F. Base.
26. B. outfeed table is slightly higher than the cutterhead
27. 8, 20
28. screw shooter
29. plate jointer

ANSWERS TO UNIT 4 QUIZ

1. True.
2. False.
3. False.
4. True.
5. False.
6. B
7. A
8. F
9. E
10. C
11. D
12. A. portable electric jig
13. D. All of the above.
14. C. 90 psi

Reproducible Master 4-1
Portable Power Circular Saw

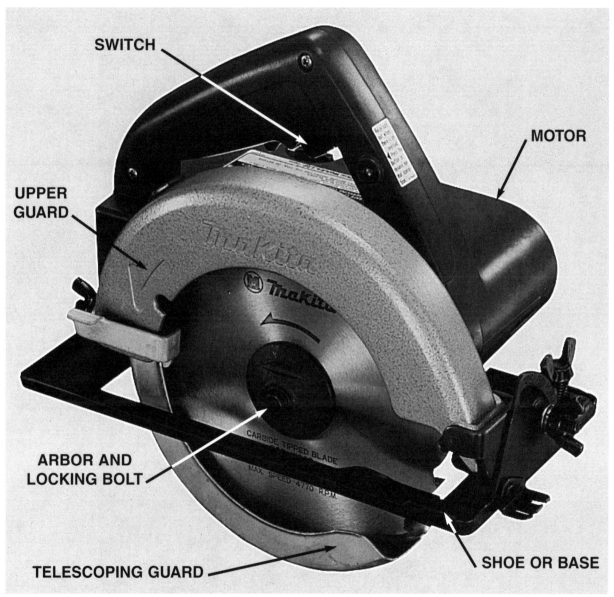

SWITCH

MOTOR

UPPER GUARD

ARBOR AND LOCKING BOLT

TELESCOPING GUARD

SHOE OR BASE

Reproducible Master 4-2
Jointer Operation

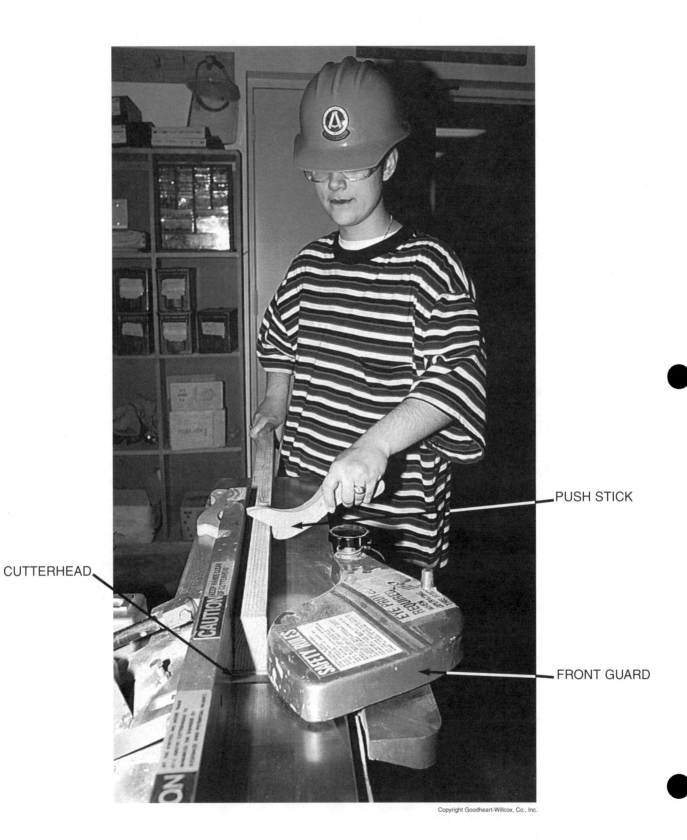

PUSH STICK

CUTTERHEAD

FRONT GUARD

Unit 4 Quiz
Power Tools

Name _____ Score _____

True-False

Circle T if the answer is True or F if the answer is False.

T   F   1.   The size of a circular saw is determined by the diameter of the largest blade it will accept.

T   F   2.   Electrical power tools operate only on 120 V of electric power.

T   F   3.   A router bit revolves in a counterclockwise direction.

T   F   4.   A power nailer that carries nails in a round canister is called a coil-fed nailer.

T   F   5.   Routers cannot be used to cut through material.

Identification

Identify the parts of the portable circular saw.

_____   6.   Motor.

_____   7.   Switch.

_____   8.   Upper guard.

_____   9.   Arbor and locking bolt.

_____   10.   Shoe or base.

_____   11.   Telescoping guard.

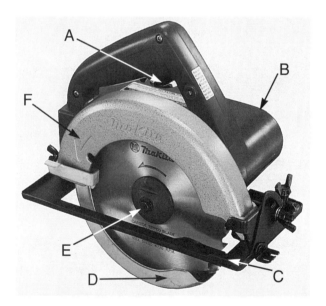

Multiple Choice

Choose the answer that correctly completes the statement. Write the corresponding letter in the space provided.

_____   12.   A saber saw is also referred to as a _____ saw.

A.   portable electric jig
B.   builder's
C.   drywall
D.   None of the above.

_____   13.   One type of portable power sander is a _____ sander.

A.   belt
B.   disc
C.   finish
D.   All of the above.

_____ 14. The air pressure for an air-powered nailer should seldom exceed _____ .
   A. 30 psi
   B. 60 psi
   C. 90 psi
   D. 120 psi

5

Leveling Instruments

OBJECTIVES

Students will be able to:

- ❏ Explain the operation of the builder's level and level-transit.
- ❏ Explain the basic operation of a laserplane system.
- ❏ Demonstrate proper setup, sighting, and leveling procedures.
- ❏ Measure and lay out angles using leveling equipment.
- ❏ Read the vernier scale.
- ❏ Use a plumb line.

INSTRUCTIONAL MATERIALS

Text: Pages 101-113
>   *Important Terms,* page 112
>   *Test Your Knowledge,* page 112
>   *Outside Assignments,* page 112

Workbook: Pages 29-32

Instructor's Manual:
>   Reproducible Master 5-1, *Transit Graduated Circle*
>   Procedure Checklist—*Setting Up and Leveling a Builder's Level*
>   Unit 5 Quiz

TRADE-RELATED MATH

When using a builder's level or level-transit, a good understanding of angular divisions is necessary. A graduated circle contains 360°. Each degree can be divided into 60 minutes. Each minute can further be divided into 60 seconds. In order to convert the number of degrees in a given number of minutes, or the number of minutes in a given number of seconds, divide the given number by 60. For example:

>   90 minutes = x degrees
>   90 ÷ 60 = 1.5
>   90 minutes = 1.5 degrees

>   120 seconds = x minutes
>   120 ÷ 60 = 2
>   120 seconds = 2 minutes

In order to find the number of minutes in a given number of degrees, or the number of seconds in a given number of minutes, multiply the given number by 60. For example:

>   5 degrees = x minutes
>   5 × 60 = 300
>   5 degrees = 300 minutes

>   8 minutes = x seconds
>   8 × 60 = 480
>   8 minutes = 480 seconds

INSTRUCTIONAL CONCEPTS AND STUDENT LEARNING EXPERIENCES

PLOT PLAN

1. Using Figure 5-1 of the text or another plot plan obtained from a contractor, identify the primary components of the plan. These items might include: length and bearing of each property line, location and size of buildings on the site, contour of the lot, elevation of the property corners, streets, drives, and other means of access, utility easements, and scale of the drawing.

ESTABLISHING BUILDING LINES

1. Stress the importance of accurately establishing the building lines on a site. Discuss possible scenarios that may occur if the building lines are not properly located.
2. Review the measurements found on a measuring tape. Have students take turns in measuring long distances and accurately reading the tape.
3. Emphasize the importance of measuring perpendicular to the boundary lines when locating the building lines. Demonstrate methods of ensuring

perpendicularity such as swinging arcs or using the 6-8-10 method.

4. Divide the students into pairs and have them establish building lines for a hypothetical plot plan. Rotate the students and have them check the measurements of the building lines established by other students.

LAYING OUT WITH LEVELING INSTRUMENTS

1. Discuss the tools that may be used to lay out building lines for small structures. Once again, stress the importance of accurately locating the building lines.

2. Identify the parts of the builder's level and level-transit. Discuss the similarities and differences between the two instruments.

3. Demonstrate the proper methods for setting up a builder's level and level-transit. Describe possible obstacles that may be encountered when setting up these instruments in different types of terrain (i.e. sandy or extremely hard and rocky soil).

4. Emphasize the importance of properly leveling the instruments when setting them up. Have students take turns leveling the instruments. Change the location of the leveling instruments after each attempt.

5. Show students the correct procedures for sighting a builder's level or level-transit over long and short distances. Describe how the leveling rod is used when sighting over long distances. Stress that both eyes should be kept open when sighting the instrument. (Most people tend to close their less-dominant eye when sighting a leveling instrument.)

6. Demonstrate the procedures for using the laser-plane system as a leveling device. If possible, invite a manufacturer's sales representative to give a demonstration.

7. Stress the importance of proper care and maintenance of all leveling systems. Emphasize the importance of following correct storage and maintenance procedures when setting up, using, and storing the instruments.

USING THE INSTRUMENTS

1. Using Reproducible Master 5-1, *Transit Graduated Circle*, define the terms commonly associated with the graduated circle. Be sure that the students fully understand these terms before progressing to reading the graduated horizontal circle of a transit.

2. Set the horizontal circle and vernier scales at dif-ferent locations on a transit. Have the students read the scales and correctly state the measurements.

3. Demonstrate the proper procedure for laying out and staking a house. As a class, lay out the building lines for a hypothetical building site. (Use 90° corners for this class exercise.) When this has been completed, divide students into pairs and have each pair lay out another building site. (Use angles other than 90° for this exercise.)

4. Define the term "grade leveling." Cite applications for grade leveling a building site.

5. Describe the means for indicating "cuts" and "fills" on a building site.

6. Define "contour lines" and have students identify contour lines on a plot plan. Ask them to interpret the meanings of several contour lines found on a plot plan.

7. Discuss the means for using a level-transit to measure vertical angles, and for laying out and checking walls.

8. Demonstrate the use of a level-transit to establish vertical angles.

UNIT REVIEW

1. Review the unit objectives. Be sure that the students fully understand each objective.

2. Assign *Important Terms, Test Your Knowledge* questions, and *Outside Assignments* on page 112 of the text. Review the answers in class.

3. Assign pages 29-32 of the workbook. Review the answers in class.

EVALUATION

1. Use Procedure Checklist—*Setting Up and Leveling a Builder's Level* to evaluate the techniques students use to perform these tasks.

2. Use Unit 5 Quiz for in-class evaluation. Correct the quizzes and return them to the students for review.

3. Using a plot plan complete with all appropriate measurements, have the students work in pairs to lay out the building lines of a structure. Emphasize that they will be evaluated based on the accuracy of their layout.

ANSWERS TO TEST YOUR KNOWLEDGE TEXT PAGE 112

1. Building lines are the lines marking where the walls of a structure are to be located on a building site. (See page 101.)

2. See page 103, Figure 5-4.

3. line of sight
4. circular
5. decimal parts of an inch
6. leveling
7. focusing knob
8. central
9. plumb bob
10. minutes, seconds

ANSWERS TO WORKBOOK QUESTIONS PAGES 29-32

1. straight
2. A. Checking plumb lines.
3. A. Focusing knob.
   B. Instrument level vial.
   C. Eyepiece.
   D. Leveling screw.
   E. Index vernier.
   F. Horizontal graduated circle.
4. C. tripod
5. B. decimal parts of a foot
6. A. 6'-3 1/4"
   B. 1.70'
7. A
8. benchmark
9. 7'-1"
10. plot
11. B. two corners
12. 47° 5 min.
13. 21,600 minutes
    54,000 seconds
14. contour lines

15. level-transit
16. B. both eyes be kept open
17. The operator places the bucket or blade cutting edge of the backhoe on the benchmark or the finished elevation. Then the receiver is adjusted up or down on the machine and the operator tightens the clamp when the "on-grade" point is reached. The receiver "catches" the laser beam from the rotating transmitter and signals the operator whether the measured surface is above, below, or on grade.

ANSWERS TO UNIT 5 QUIZ

1. False.
2. False.
3. True.
4. True.
5. D. 90°
6. A. left thumb moves
7. B. 60
8. C. grade leveling
9. G
10. H
11. C
12. D
13. F
14. I
15. E
16. B
17. A

Reproducible Master 5-1
Transit Graduated Circle

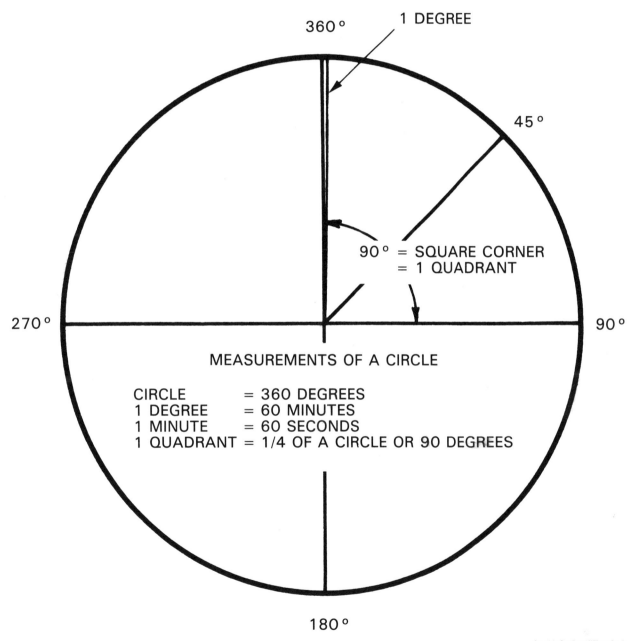

360°   1 DEGREE

45°

90° = SQUARE CORNER
= 1 QUADRANT

270°   90°

MEASUREMENTS OF A CIRCLE

CIRCLE   = 360 DEGREES
1 DEGREE   = 60 MINUTES
1 MINUTE   = 60 SECONDS
1 QUADRANT = 1/4 OF A CIRCLE OR 90 DEGREES

180°

Unit 5 Procedure Checklist
Setting Up and Leveling a Builder's Level

Name _____ Total _____

❏ Properly sets up tripod; makes sure points are secure; properly adjusts tripod legs.	5	4	3	2	1	Treats tripod with little respect; legs are unstable; narrow box formed by tripod legs.
❏ Properly mounts instrument on tripod, taking care not to damage it.	5	4	3	2	1	Reckless use of instrument; damage of instrument.
❏ Correctly levels the instrument in one direction using both thumbs to adjust levelness.	5	4	3	2	1	Cannot level instrument; unable to move the level bubble toward center of vial.
❏ Turns level 90° over other set of leveling screws; properly adjusts levelness.	5	4	3	2	1	Neglects to level instrument in other direction.
❏ Rotates level in complete circle and checks for bubble movement.	5	4	3	2	1	Neglects to verify levelness.

Unit 5 Quiz
Leveling Instruments

Name _____ Period _____

True-False

Circle T if the answer is True or F if the answer is False.

T F 1. Building lines are the lines indicating where the footings will be constructed.

T F 2. The laser plane receiver must always be tended by a rod person.

T F 3. Both eyes should be kept open when sighting a leveling instrument.

T F 4. A reference point where a leveling instrument is located is called a station point.

Multiple Choice

Choose the answer that correctly completes the statement. Write the corresponding letter in the space provided.

_____ 5. The 6-8-10 method can be used to verify _____ corners.

        A. 30°
        B. 45°
        C. 60°
        D. 90°

_____ 6. When adjusting a builder's level for levelness, the bubble in the level vial will travel in the direction that your _____.

        A. left thumb moves
        B. right thumb moves
        C. telescope is pointed
        D. None of the above.

_____ 7. One degree contains _____ minutes.

        A. 30
        B. 60
        C. 90
        D. 120

_____ 8. Finding the difference in the grade level between several points is called _____.

        A. grading
        B. calibrating
        C. grade leveling
        D. grade staking

Identification

Identify the parts of the builder's level.

_____ 9. Tripod mounting stud.

_____ 10. Leveling screws.

_____ 11. Instrument level vial.

_____ 12. Eyepiece.

_____ 13. Horizontal clamp screw.

_____ 14. Horizontal graduated circle.

_____ 15. Index vernier.

_____ 16. Focusing knob.

_____ 17. Telescope lens.

6

Plans, Specifications, and Codes

OBJECTIVES

Students will be able to:

- ❏ Identify the elements commonly included in a set of house plans.
- ❏ Demonstrate the use of scale in architectural drawings.
- ❏ Recognize architectural symbols.
- ❏ Explain the use of building specifications.
- ❏ Summarize the concept of modular construction.
- ❏ Describe the application of building codes, standards, and permits.

INSTRUCTIONAL MATERIALS

Text: Pages 114-139
    Important Terms, page 137
    Test Your Knowledge, page 138
    Outside Assignments, page 138
Workbook: Pages 33-36
Instructor's Manual:
    Reproducible Master 6-1, *Architectural Symbols, Parts A and B*
    Reproducible Master 6-2, *Symbols for Openings*
    Reproducible Master 6-3, *Appliance/Fixture Symbols*
    Procedure Checklist—*Reading a Set of Working Drawings*
    Unit 6 Quiz
    Section I Exam

TRADE-RELATED MATH

Foot and inch values are the most common type of measurement used in construction. In many instances, foot and inch values must be added or subtracted to obtain a final dimension. When performing any type of calculation using foot and inch values, treat the foot value and inch value separately. Then, combine the results to obtain the final answer.

When adding, if the sum (result) of the inch value computation is more than 12″, convert the sum to a foot value by dividing by 12 and using the remainder as the new inch value. For example:

$$6'\text{-}9'' + 2'\text{-}4'' + 1'\text{-}6'' = x$$

$$
\begin{array}{r}
6'\text{-}\ 9'' \\
2'\text{-}\ 4'' \\
+\ 1'\text{-}\ 6'' \\
\hline
9'\text{-}19'' = 10'\text{-}7''
\end{array}
$$

$$x = 10'\text{-}7''$$

When subtracting foot and inch values, you may need to borrow 12″ (1′) from the left column. Then subtract as you would any other whole numbers. For example:

$$7'\text{-}6'' - 4'\text{-}9'' = x$$

Since 9″ cannot be subtracted from 6″, 12″ is borrowed from the foot value. The 7′-6″ value is equal to 6′-18″. Now, subtract the two values.

$$
\begin{array}{r}
6'\text{-}18'' \\
4'\text{-}\ 9'' \\
\hline
2'\text{-}\ 9''
\end{array}
$$

$$x = 2'\text{-}9''$$

INSTRUCTIONAL CONCEPTS AND STUDENT LEARNING EXPERIENCES

SET OF PLANS

1. Using Figures 6-5 through 6-17 or another complete set of plans, review the primary drawings, including: the plot plan, the foundation plan, the floor plans, the elevations drawings, and the mechanical and electrical layouts.

2. Discuss the reasons for scaling drawings for structures. Explain what notations such as

1" = 1'-0" represent on a plan. Discuss why a 1/4" = 1'-0" scale is commonly used for residential plans.

3. Review the purpose of a plot plan. Ask the students to list the major components of the plan.

4. Using Figure 6-5 (or another floor plan), Reproducible Master 6-1, *Architectural Symbols, Parts A* and *B*, Reproducible Master 6-2, *Symbols for Openings*, and Reproducible Master 6-3, *Appliance/Fixture Symbols*, have students identify the symbols commonly used on floor plans. Discuss the importance of readily identifying and interpreting symbols in becoming a productive carpenter.

5. Have the students develop a floor plan for a simple, one-story structure. Emphasize the use of appropriate symbols on the floor plan. Ask the students to write a thorough description of the floor plan using only words. Discuss the advantages of using symbols on plans rather than a written description.

6. Using Figures 6-5 and 6-6 (or another compatible floor plan and foundation plan), ask the students to locate features that are common to both plans such as the stairs, building walls, etc.

7. Review the elevation drawings shown in Figures 6-8 and 6-9 (or another comparable set of elevations). Ask the students to identify any symbols they recognize. Using Reproducible Master 6-1, *Architectural Symbols,* Parts *A* and *B*, compare the symbols used on floor plans and elevations.

8. Describe the relationship between floor plans and elevations. Ask students to correlate features shown on the floor plan with their respective symbols on an elevation. Stress that features shown on an elevation will likely be located by dimensions given on a floor plan.

9. Using Figures 6-10 through 6-13, discuss the use of framing plans for a house. Ask the students to describe how the placement of framing members will be determined if framing plans are not included with the set of plans. Discuss where the dimensions will be found that can help them determine the placement of the framing members if framing plans are not included.

10. Discuss the importance of using sections and detail drawings to provide additional information about the construction of a house. Using Figures 6-14 and 6-18 and Reproducible Master 6-1, *Architectural Symbols, Parts A* and *B*, ask the students to interpret the sections and details.

11. Using the set of plans that were used for the previous discussions, explain the importance of accurately reading and interpreting dimensions on a drawing. Stress that dimensions greater than one foot are usually shown in feet and inches, such as 1'-8", rather than 20".

12. Review the Trade-Related Math concepts to illustrate the addition and subtraction of dimensions. Using the floor plan of a structure, add and/or subtract the overall dimensions with other feature dimensions to verify accuracy.

13. Review the items commonly found on a materials list, including quantity, name of the material, description, and size. Have the students develop an abbreviated materials list for the simple, one-story structure they developed in item 5.

HOW TO SCALE A DRAWING

1. Discuss the means for determining missing dimensions on a drawing including using an architect's scale or a folding rule to scale a plan.

2. Hand out architect's scales to the class. Ask students to draw lines of varying lengths and measure each line using the 1/4" = 1'-0" and 1/8" = 1'-0" scales. Place the measurements next to each line. Have the students exchange their drawings with each other for verification.

CHANGING PLANS

1. Explain the ramifications of changes made in plans. Discuss possible scenarios that may occur if plans are changed without notifying other tradesworkers.

SPECIFICATIONS

1. Explain the purpose of developing a list of specifications in addition to a set of plans. Have the students list the headings generally included in the specifications for a residential structure. Discuss the importance of having a thorough set of specifications.

MODULAR CONSTRUCTION

1. Discuss the concept of modular construction. Emphasize that many modern construction materials are based on the modular system including plywood panels and concrete blocks. Explain the advantages of using modular construction.

2. Define the terms "minor module" and "major module."

3. Discuss how the modular construction concept applies to the SI metric system.

BUILDING CODES

1. Describe the purpose of building codes. Obtain a copy of the local building code and discuss the importance of conforming to the code.
2. Obtain a variety of current model codes including the Uniform Building Code, the BOCA—Basic Building Code, the National Building Code, and the Standard Building Code. Have the students locate similar items in the codes and compare the requirements of each code.
3. Assign a particular topic that is addressed in all of the major model codes. Ask the students to research the topic using at least two of the model codes and the local building code as references.
4. Discuss the organizations and associations that play a key role in the development of model codes. Ask the students to explain why they think these organizations or associations are involved in the development of standards.
5. Invite a local building inspector to discuss the process of obtaining a building permit in your community. Allow students to ask questions related to the building permit and inspection process.

UNIT REVIEW

1. Review the unit objectives. Be sure that the students fully understand each objective.
2. Assign *Important Terms, Test Your Knowledge* questions, and *Outside Assignments* on pages 137 and 138 of the text. Review the answers in class.
3. Assign pages 33-38 of the workbook. Review the answers in class.

EVALUATION

1. Using the Procedure Checklist—*Reading a Set of Working Drawings*, have the students identify symbols, perform math computations for missing dimensions, and interpret the working drawings of a residential structure.
2. Use Unit 6 Quiz for in-class evaluation. Correct the quizzes and return them to the students for review.
3. Use Section I Exam to evaluate the students' knowledge of information found in Units 1-6 of the text.

ANSWERS TO TEST YOUR KNOWLEDGE TEXT PAGE 138

1. Plot plan; foundation or basement plan; floor plans; elevation drawings; drawings of electrical, plumbing, heating, and air conditioning layouts.
2. $1/4'' = 1'-0''$.
3. size
4. location of the structure on the building site
5. outside
6. part, cut by a vertical plane
7. distances
8. See text, pages 129-131.
9. architect's
10. specifications, specs.
11. 4
12. Computer Assisted Drafting and Design
13. 300 millimeters
14. True.

ANSWERS TO WORKBOOK QUESTIONS PAGES 33-38

1. A. 48'-4"
   B. 8"
   C. 16" × 8"
   D. 24" × 24" × 12"
   E. 7
   F. 48'-0"
   G. 2" × 8"
   H. 12" O.C.
2. A. 28'-8"
   B. 6'-8"
   C. 3
   D. 15'-5"
   E. 20'-2"
   F. 9'-1 1/2"
   G. Linoleum.
   H. 2" × 6"
   I. 16" O.C.
   J. 20" × 20"
   K. Sliding (pocket).
   L. Casement.
   M. 1'-8"
   N. One.
   O. 4"
   P. Hardwood (oak).
3. A. Wood.
   B. Brick.

C. Stone.
D. Concrete.
E. Concrete block.
F. Earth.
G. Glass.
4. 38′-2″
5. A. Double hung window.
   B. Casement window.
   C. Sliding doors (exterior).
   D. Bifold doors.
   E. Pocket sliding door.
   F. Arch.
   G. Bypass sliding door.
   H. Door (swinging).
6. A. Shower (built-in).
   B. Lavatory (built-in).
   C. Stool.
   D. Tub (recessed).
   E. Refrigerator.
   F. Cooking top.
   G. Kitchen sink.
   H. Range.
7. D. Number of painters to be used on the job.
8. trade associations
9. A. Commerce.
   B. CS.
10. owner
11. C. inspector

ANSWERS TO UNIT 6 QUIZ

1. False.
2. True.
3. True.
4. True.
5. False.
6. B. Floor plans
7. C. erasures
8. D. Room size.
9. B. 4″
10. D
11. E
12. F
13. C
14. B
15. A

ANSWERS TO SECTION 1 EXAM

1. False.
2. True.
3. True.
4. False.
5. True.
6. True.
7. C. studs, joists, rafters, and wall plates
8. D. 12%
9. B. face
10. A. width of the belt
11. B. 6″
12. D. 12″
13. D. cambium layer
14. E
15. A
16. C
17. B
18. D
19. F
20. G
21. Eight
22. plot
23. 45
24. line of sight
25. benchmark
26. contour
27. 1 1/2 × 3 1/2
28. 120
29. contact cement
30. 10
31. 8
32. 1/8
33. D
34. B
35. A
36. G
37. E
38. F
39. C
40. E
41. A
42. C
43. F
44. B
45. D

Reproducible Master 6-1 (Part A)
Architectural Symbols

Material	Plan	Elevation	Section
Wood	FLOOR AREAS LEFT BLANK	SIDING PANEL	FRAMING FINISH
Brick	FACE COMMON	FACE OR COMMON	SAME AS PLAN VIEW
Stone	CUT RUBBLE	CUT RUBBLE	CUT RUBBLE
Concrete			SAME AS PLAN VIEW
Concrete Block			SAME AS PLAN VIEW
Earth	NONE	NONE	

Reproducible Master 6-1 (Part B)
Architectural Symbols

Material	Plan	Elevation	Section
Glass			LARGE SCALE / SMALL SCALE
Insulation	SAME AS SECTION	INSULATION	LOOSE FILL OR BATT / BOARD
Plaster	SAME AS SECTION	PLASTER	STUD / LATH AND PLASTER
Structural Steel	INDICATE BY NOTE	INDICATE BY NOTE	
Sheet Metal Flashing	INDICATE BY NOTE		SHOW CONTOUR
Tile	FLOOR	WALL	

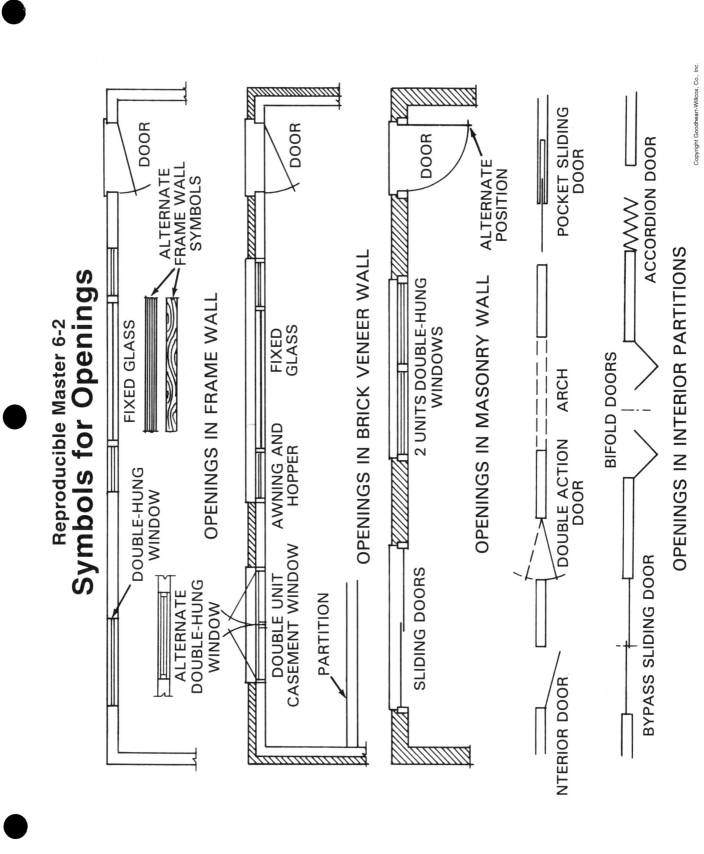

Reproducible Master 6-2
Symbols for Openings

DOUBLE-HUNG WINDOW

FIXED GLASS

DOOR

ALTERNATE FRAME WALL SYMBOLS

OPENINGS IN FRAME WALL

ALTERNATE DOUBLE-HUNG WINDOW

DOUBLE UNIT CASEMENT WINDOW

AWNING AND HOPPER

FIXED GLASS

DOOR

PARTITION

OPENINGS IN BRICK VENEER WALL

SLIDING DOORS

2 UNITS DOUBLE-HUNG WINDOWS

DOOR

ALTERNATE POSITION

OPENINGS IN MASONRY WALL

POCKET SLIDING DOOR

ARCH

DOUBLE ACTION DOOR

BIFOLD DOORS

ACCORDION DOOR

INTERIOR DOOR

BYPASS SLIDING DOOR

OPENINGS IN INTERIOR PARTITIONS

Reproducible Master 6-3
Appliance/Fixture Symbols

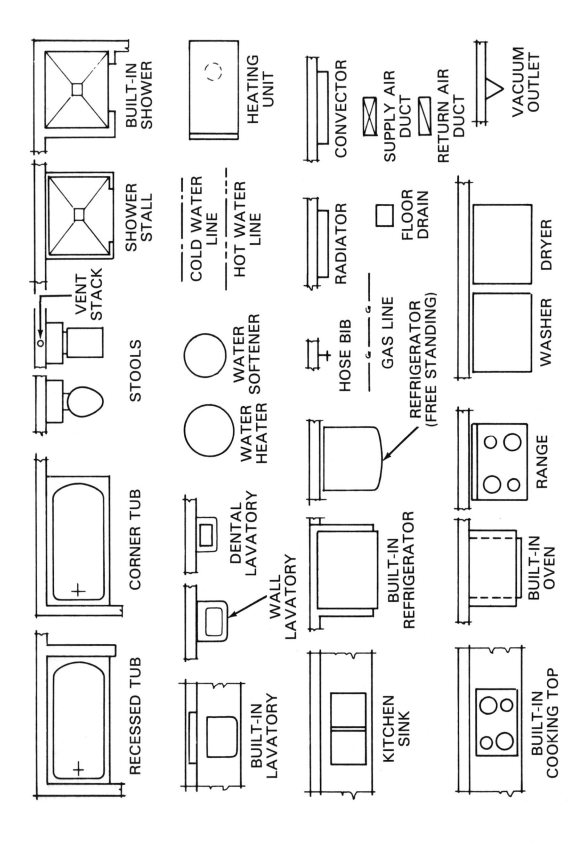

BUILT-IN SHOWER

HEATING UNIT

CONVECTOR

SUPPLY AIR DUCT

RETURN AIR DUCT

VACUUM OUTLET

SHOWER STALL

COLD WATER LINE

HOT WATER LINE

RADIATOR

FLOOR DRAIN

DRYER

VENT STACK

STOOLS

WATER SOFTENER

WATER HEATER

HOSE BIB

GAS LINE

REFRIGERATOR (FREE STANDING)

WASHER

CORNER TUB

DENTAL LAVATORY

WALL LAVATORY

BUILT-IN REFRIGERATOR

RANGE

RECESSED TUB

BUILT-IN LAVATORY

KITCHEN SINK

BUILT-IN OVEN

BUILT-IN COOKING TOP

Unit 6 Procedure Checklist
Reading a Set of Working Drawings

Name _____ Total _____

❏ Identifies the scale of a drawing and explains its meaning.	5	4	3	2	1	Unable to identify the scale; cannot explain the importance of a scale on a drawing.
❏ Readily identifies drawings in a set without reading the title; explains the purpose of each.	5	4	3	2	1	Unable to differentiate between types of drawings in a set; cannot explain the purpose of the drawings.
❏ Interprets drawings when instructor requests dimensions for particular feature of the plan.	5	4	3	2	1	Cannot identify drawing containing appropriate dimensions; unable to correctly read dimensions for particular features.
❏ Readily identifies symbols of materials and other architectural items.	5	4	3	2	1	Unable to identify symbols when requested by instructor.
❏ Scales dimensions on a scaled drawing.	5	4	3	2	1	Cannot read an architect's scale.
❏ Reads and understands specifications included with a set of drawings; states the purpose of specifications.	5	4	3	2	1	Cannot locate specifications in a set; unable to state purpose of specifications.

Unit 6 Quiz

Plans, Specifications, and Codes

Name _____ Score _____

True-False

Circle T if the answer is True or F if the answer is False.

T F   1. The window and door schedule states when the windows and doors will be delivered to the construction site.

T F   2. The Unicom system makes it possible to apply mass production methods to building construction.

T F   3. Dimension lines on architectural drawings are continuous lines with the dimensions placed above them.

T F   4. A line measuring 4'-0" on a drawing with a 1/4" = 1'-0" scale will appear larger than the same line on a 1/8" = 1'-0" drawing.

T F   5. The modular system in the SI metric system is based on a grid made of 100 meter squares.

Multiple Choice

Choose the answer that correctly completes the statement. Write the corresponding letter in the space provided.

_____ 6. _____ show the size and outline of a building and its rooms.

     A. Elevation drawings
     B. Floor plans
     C. Plot plans
     D. Section drawings

_____ 7. A major advantage of CADD is that a drawing can be changed without making _____ on drafting paper.

     A. symbols
     B. notes
     C. erasures
     D. corrections

_____ 8. Which of the following *cannot* be determined from an elevation drawing?

     A. Floor level.
     B. Window and door heights.
     C. Roof slopes.
     D. Room size.

_____ 9. All dimensions in modular construction are based on multiples of _____ .

     A. 3"
     B. 4"
     C. 5"
     D. 6"

Identification

Identify the following window and door symbols.

_____ 10. Double hung window.

_____ 11. Fixed glass window.

_____ 12. Sliding doors.

_____ 13. Double action door.

_____ 14. Pocket door.

_____ 15. Bifold doors.

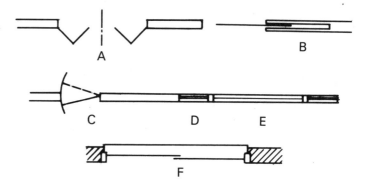

Section 1 Exam
Preparing to Build

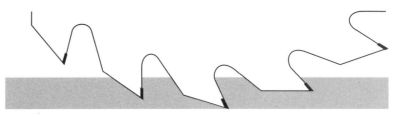

Name _____ Score _____

True-False

Circle T if the answer is True or F if the answer is False.

T F   1. Open grain woods generally do not require additional preparation during finishing.

T F   2. The best grade of softwood lumber is Select.

T F   3. Engineered lumber does not warp; however, it is not as strong as sawed lumber.

T F   4. Nominal dimensions of lumber are always greater than the dressed dimensions.

T F   5. A small backsaw should be used to cut curves.

T F   6. The blade rotation of a radial arm saw is in the direction of the saw feed.

Multiple Choice

Choose the answer that correctly completes the statement. Write the corresponding letter in the space provided.

_____   7. Steel structural materials are available in several gauges and often replace wood _____ .
   A. subflooring
   B. bracing
   C. studs, joists, rafters, and wall plates
   D. trim

_____   8. The moisture content of a pieces of wood weighing 45 lb. before it was dried and 40 lb. after it was dried is _____ .
   A. 1%
   B. 5%
   C. 7%
   D. 12%

_____   9. The _____ of a hammer should strike the nail head when driving a nail.
   A. cheek
   B. face
   C. neck
   D. handle

_____ 10. The size of a belt sander is determined by the _____.
   A. width of the belt
   B. length of the belt
   C. size of the sander motor
   D. None of the above.

_____ 11. When using a radial arm saw, a _____ margin of safety should be maintained between the operator's hands and the path of the saw blade.

    A. 1″

    B. 6″

    C. 1′

    D. 1′-6″

_____ 12. When using a jointer, the stock must be no less than _____ long.

    A. 2″

    B. 4″

    C. 8″

    D. 12″

_____ 13. Wood cells are formed in the _____.

    A. tracheids

    B. phloem

    C. lignen

    D. cambium layer

Matching

Select the correct answer from the list on the right and place the corresponding letter in the blank on the left.

_____ 14. Laminated veneer lumber.

_____ 15. Glue-laminated beams.

_____ 16. Wood I-beams.

_____ 17. Open-web trusses.

_____ 18. Parallel-strand lumber.

_____ 19. Laminated-strand lumber.

_____ 20. Steel structural members.

    A. Laminated from 2 × 4 or 2 × 6 lumber to form beams.

    B. Fabricated of solid 2 × 4 chords with webbing of steel.

    C. Made of solid lumber chords between a web of plywood or oriented strand board.

    D. Veneer layed up parallel and glued under pressure.

    E. Manufactured in billets 66′ long and cut to various lengths and widths.

    F. Layed up from 1/32 × 1 × 12″ strands bonded with polyurethane adhesive.

    G. Manufactured with flanges for greater rigidity.

Completion

Place the answer that correctly completes the statement in the space provided.

_____ 21. _____ board feet of lumber is contained in a piece measuring 2″ × 12″ × 48″.

_____ 22. A _____ plan indicates the location of the structure and the distances from it to the property lines.

_____ 23. One quadrant contains _____ degrees.

_____ 24. The operating principle of a leveling instrument is that a _____ is always a straight line.

_____ 25. The officially established elevation that can be used for a number of building sites in a given location is called the _____.

Name _____

_____ 26. The lines on a map that run through points of equal elevation are _____ lines.

_____ 27. The dressed dimensions of a 2 × 4 piece of lumber to be used as a stud is _____.

_____ 28. Ten 2 × 6s that are 12′ long contain a total of _____ board feet.

_____ 29. Plastic laminates are usually bonded to a base material with _____.

_____ 30. A(n) _____d nail is 3″ long.

_____ 31. A 1/2″ auger bit will have a number _____ stamped on its tang.

_____ 32. The blade of a portable circular saw should be adjusted so that it cuts through the stock and projects about _____ ″.

Identification

Identify the parts of the log cross section.

_____ 33. Cambium.

_____ 34. Sapwood.

_____ 35. Heartwood.

_____ 36. Pith.

_____ 37. Bark.

_____ 38. Annual rings.

_____ 39. Wood rays.

Identification

Identify the parts of the grade-trademark.

_____ 40. Thickness.

_____ 41. Panel grade.

_____ 42. Exposure durability classification.

_____ 43. Mill number.

_____ 44. Span rating.

_____ 45. National Research Board report number.

A —— RATED SHEATHING
B —**32/16** 15/32 INCH —— E
SIZED FOR SPACING
C ——— EXPOSURE 1
=000= — F
NER-QA397 PRP-108
D

7

Footings and Foundations

OBJECTIVES

Students will be able to:

- ❏ Lay out building lines and set up batter boards.
- ❏ Describe excavation procedures.
- ❏ Explain footing requirements and how to build footing forms.
- ❏ Define the terms "concrete," "cement," and "aggregate."
- ❏ Describe the building, erecting, and use of forms for poured foundation walls.
- ❏ Demonstrate techniques for forming openings in poured foundation walls.
- ❏ List steps and professional practices for laying up concrete block foundation walls.
- ❏ Explain foundation insulating and waterproofing procedures.
- ❏ Discuss design factors that apply to sidewalks and driveways.
- ❏ Estimate concrete materials required for a specific area.

INSTRUCTIONAL MATERIALS

Text: Pages 143-178

　　Important Terms, page 176
　　Test Your Knowledge, page 176
　　Outside Assignments, page 177

Workbook: Pages 39-46

Instructor's Manual:

　　Reproducible Master 7-1, *Using Batter Boards, Parts A* and *B*
　　Reproducible Master 7-2, *Foundation Elements*
　　Reproducible Master 7-3, *Foundation Types*
　　Reproducible Master 7-4, *Laying Out Footing Forms*
　　Reproducible Master 7-5, *Locating Footing Forms*
　　Reproducible Master 7-6, *Concrete Blocks*
　　Reproducible Master 7-7, *Wood Foundation Design*

Procedure Checklist—*Setting Up Batter Boards*
Procedure Checklist—*Constructing Footing Forms*
Unit 7 Quiz

TRADE-RELATED MATH

Concrete is measured and sold by the cubic yard. A cubic yard measures $3' \times 3' \times 3'$. It contains 27 cu. ft. ($3 \times 3 \times 3 = 27$). To determine the number of cubic yards needed for a square or rectangular area, multiply the width by the length by the height (or thickness), and then divide by 27. For example, if you want to determine the number of cubic yards of concrete needed for a concrete slab measuring $20' \times 15' \times 4''$ (.33') thick, calculate it as follows:

$$\frac{\text{width} \times \text{length} \times \text{thickness}}{27} = \text{cubic yards}$$

$$\frac{20 \times 15 \times .33}{27} = \frac{99}{27} = 3.66$$

3.66 cubic yards of concrete are required

Tables are available for determining the number of concrete blocks and the bags per 100 sq. ft. of wall. Refer to the table, "Estimating Concrete Block and Mortar" in the Technical Information section of the text.

Suppose that you wish to determine the amount of block and cement needed for a basement 8' high and 24' square, using standard $8 \times 8 \times 16''$ block. The chart, "Number of Block/Course for Solid Walls," in the text's Technical Information Section shows that 70 block are needed for each course. According to the chart, "Concrete Masonry Courses by Heights," also found in the Technical Information section, shows that 12 courses are needed. Thus:

Number of Block in a Course × Number of Courses = Number of Blocks Needed = 70 ´ 12 = 840.

Number of blocks needed: 840; however, since there is some waste, estimators usually increase this total by 2%:

840 × 1.02 = 856.8 which is rounded off to the next larger whole number, 857 blocks needed.

INSTRUCTIONAL CONCEPTS AND STUDENT LEARNING EXPERIENCES

CLEARING THE SITE
1. Describe typical operations performed when clearing a building site. Discuss the advantages of retaining conifers and deciduous trees on the property when clearing the site.

LAYING OUT BUILDING LINES
1. Discuss why a registered engineer or licensed surveyor should verify the position of the lot lines.
2. Review the methods of laying out the building lines. (This was discussed in Unit 5 of the text.) Describe the means of checking for squareness of building lines.
3. Explain the purpose of setting up batter boards. Using parts *A* and *B* of Reproducible Master 7-1, *Using Batter Boards,* have the students list the materials necessary used to construct batter boards.
4. Given the floor plan of a residential structure, have the students indicate where they would position the batter boards.
5. Demonstrate the correct procedure for setting up batter boards. As a class, lay out the building lines for a small residential structure and construct batter boards. Emphasize the importance of constructing the batter boards accurately.

EXCAVATION
1. Describe the different operations that may occur when excavating the building site.
2. Discuss why foundations must be located below the frost line in cold climates. Explain what might happen if a foundation is not located deep enough.
3. Define the term "control point."
4. Describe factors that should be considered when determining the depth of an excavation.

FOUNDATION SYSTEMS
1. Using Reproducible Master 7-2, *Foundation Elements,* identify the components of a foundation. Ask the students to briefly describe the purpose of each component.
2. Using Reproducible Master 7-3, *Types of Foundations*, discuss the various types of foundations used in residential construction.
3. Refer to Figure 7-10. Have the students identify the different types of footings. Compare and contrast each of these types of footings and describe situations where each might be used.
4. Discuss the different types of slab foundations including structurally supported, ground supported, and monolithic. Describe applications and advantages of each type.

FOOTING DESIGN
1. Discuss the typical dimensions for standard footings, footings located under columns and posts, and chimney footings.
2. Describe the applications of reinforced footings in residential construction. Identify the types of materials commonly used for reinforcement.
3. Explain the purpose of forming a key in the footings when cast-in-place concrete walls are to be used. Describe how the key is formed in the concrete.

FORMS FOR FOOTINGS
1. Discuss the importance of checking the placement of the batter boards after the excavation.
2. Using Reproducible Master 7-4, *Laying Out Footing Forms,* describe the proper procedure for laying out footing forms. Stress the importance of accurate measurements and layout techniques. Identify the types of materials that are commonly used to construct footing forms. Emphasize the importance of having an adequate excavation slope to avoid the earthen wall from collapsing.
3. Review Reproducible Master 7-5, *Locating Footing Forms.* Demonstrate the construction of footing forms for a residential structure. Have students assist you in laying out the proper position of the forms and actually constructing the forms. When this has been completed, divide the students into pairs, and have each pair lay out and construct a small section of footing forms. Use nontypical dimensions for the footings so that students must make adjustments to the width and height of the forms.

CONCRETE
1. Identify the primary components of concrete and briefly describe each. Discuss the importance of

using good quality materials and being able to recognize unsuitable materials.

2. Obtain samples of different types of aggregate. Have the students list characteristics of each.

3. Demonstrate the proper techniques for mixing concrete. Emphasize the correct order of adding materials to the mix.

ERECTING WALL FORMS

1. Identify the different types of wall forms that are used in residential construction, including site built and prefabricated forms. Discuss the advantages of each type.

2. Calculate the amount of pressure that is created by concrete at the bottom of wall forms. Compare the amount of pressure that is created at the bottom of a 4′ wall versus a 10′ wall. Stress the importance of using materials that can withstand this pressure.

3. Discuss the procedure for setting up foundation wall forms. Emphasize that correct placement of wall forms is vital to the remainder of the building construction.

4. Obtain a variety of form hardware including snap ties, taper ties, coil ties, and corner clamps. Allow the students to inspect the hardware.

5. As a class, have the students construct a scale model of a site-built wall form. Be sure to include all bracing and form hardware. Stress the importance of carefully checking the dimensions of the formwork prior to pouring concrete.

6. Discuss the advantages of using panel forms. Explain that panel forms are designed to be used many times, and thus should be removed and handled carefully to avoid damaging them.

7. Discuss the methods used to form frame openings in foundation walls.

8. Using the scale model of the site-built wall form, add formwork for a window opening. Once again, stress the importance of accurate placement of the formwork.

9. Define the term "pilaster." Describe situations where pilasters may be necessary.

PLACING CONCRETE

1. Describe methods commonly used to deliver concrete to the forms such as ready-mix trucks, concrete buckets, concrete buggies, and wheelbarrows.

2. Define "segregation" and list causes of it.

3. Demonstrate methods used to vibrate and compact concrete in forms. Stress that over-vibration creates additional pressure in the forms and causes segregation to occur.

4. Identify different types of anchors used to secure wall plates to the top of a concrete or concrete block foundation wall.

5. Using the scale model of the site-built wall form, demonstrate the correct procedure for placing concrete and vibrating it. When the placement of the concrete is complete, set anchor bolts to accept a 2 × 4 wall plate.

CONCRETE BLOCK FOUNDATIONS

1. Discuss the composition of concrete blocks.

2. Using Reproducible Master 7-6, *Concrete Blocks,* identify the various types of concrete blocks and cite applications for each. Have the students take note that the actual size of a block is 3/8″ less than the nominal size to allow for a 3/8″ mortar joint.

3. Discuss the materials used in mortar and their important properties that create a strong bond for the blocks.

4. Explain procedures for laying block. Discuss the proper tools for getting a level, plumb wall.

5. Explain the use and installation of anchors.

6. Define the term "lintel." Discuss how lintels play an important part in supporting masonry across the tops of door and window openings. Identify various techniques used to form lintels.

INSULATING FOUNDATION WALLS

1. Describe the means of insulating concrete block walls. Cite examples of where one method or another might be advantageous.

2. Explain why waterproofing is necessary. Discuss the methods used to waterproof concrete and concrete block walls.

3. Demonstrate the proper procedure for waterproofing concrete and concrete block foundation walls. Allow students to assist you in applying the waterproofing materials.

BACKFILLING

1. Stress the importance of bracing foundation walls before backfilling.

SLAB-ON-GRADE CONSTRUCTION

1. Using Figure 7-57, describe the three types of slab-on-grade foundations. List applications for each type.

2. Describe preparations that must be made before placing a slab-on-grade. Emphasize the importance of insulation and moisture control when creating the subgrade.

3. Discuss the use of perimeter insulation for slab-on-grade construction in cold climates.

BASEMENT FLOORS

1. Describe where in the construction sequence concrete is placed for a basement floor. Discuss the type of preparation that is necessary before placing the concrete.

ENTRANCE PLATFORMS AND STEPS

1. Describe the construction of formwork for steps. Cite obstacles that may be encountered when constructing the forms, such as building forms between two existing walls.
2. Demonstrate the construction of the formwork for steps. Stress the importance of adequate bracing.

SIDEWALKS AND DRIVES

1. Have students summarize the general dimensions for walks. Discuss the materials required to construct the formwork.
2. Describe how a piece of asphalt impregnated composition board is placed between a walk and other structure such as foundation walls or entrance platforms. Ask the students to explain why this is necessary.
3. Discuss the general dimensions of driveways. Have students contrast the differences between the dimensions for walks and driveways. Ask them to explain why they think these differences occur.
4. Define the terms "screeding" and "floating." Discuss the proper procedure for screeding and floating a walk or driveway.
5. Explain the purpose of using control joints in a walk or driveway. Discuss how control joints are made.
6. Describe the final steps that should be taken to finish a concrete walk or driveway (including edging and brushing). In addition, discuss how concrete should be protected against extreme heat, rain, and cold.

WOOD FOUNDATIONS

1. Describe why wood foundations have obtained a great deal of popularity. Discuss the type of preparation that must be made to the wood components.
2. Explain the type of preparation that must be made to the grade before constructing an all-weather wood foundation.
3. Using Reproducible Master 7-7, *Wood Foundation Design,* have the students list the types of materials that might be used in constructing an all-weather wood foundation.

COLD WEATHER CONSTRUCTION

1. Discuss the methods used to protect concrete in cold weather.
2. Cite techniques that might be used to ensure the integrity of concrete and masonry materials.
3. Define "admixture." Describe how admixtures can be used to change the working properties of concrete.

ESTIMATING MATERIALS

1. Discuss the method used to determine the amount of concrete needed for a particular project. Make sure that an extra 5% to 10% is added for waste.
2. Discuss the method used to determine the number of concrete masonry units (concrete blocks) needed for a particular construction project. Be sure to add 2% for waste.

UNIT REVIEW

1. Review the unit objectives. Be sure that the students fully understand each objective.
2. Assign *Important Terms, Test Your Knowledge* questions, and *Outside Assignments* on pages 176-177 of the text. Review the answers in class.
3. Assign pages 39-46 of the workbook. Review the answers in class.

EVALUATION

1. Use Procedure Checklist—*Setting Up Batter Boards,* to evaluate the techniques that students used to perform these tasks.
2. Use Procedure Checklist—*Constructing Footing Forms,* to evaluate the techniques that students used to perform these tasks.
3. Use Unit 7 Quiz for in-class evaluation. Correct the quizzes and return them to the students for review.

ANSWERS TO TEST YOUR KNOWLEDGE TEXT PAGES 176-177

1. sometimes
2. False.
3. 4'
4. True.
5. frost
6. twice
7. stepped
8. See Figure 7-12. Thickened reinforced portion of a slab foundation.
9. 2' to 3'
10. should

11. cement, sand, gravel
12. hydration
13. 94
14. cubic yard
15. Thickened portion of a concrete or masonry wall. (See Figure 7-32.)
16. smooth
17. anchors, bolts, straps
18. Lay out the blocks "dry" and adjust them for the least amount of cutting.
19. To establish level for each course. Then a line can be stretched from corner to corner as a guide for the laying of blocks.
20. Use open core blocks and fill with concrete, or: use cap blocks which have a solid top.
21. To level surface by removing the excess concrete.
22. Makes a smoother finish, fills hollows, and compacts the concrete.
23. Will bring too many fines to the surface; this will produce fine cracks in the cured concrete.
24. 7 5/8 × 7 5/8 × 15 5/8
25. cement plaster, bituminous or asphaltic material, or polyethylene film
26. Unreinforced slab with footing and wall; reinforced slab; monolithic slab. (See Figure 7-57.)
27. vapor barrier, water
28. 4″
29. porous gravel, 4″ to 6″
30. False.
31. 113

ANSWERS TO WORKBOOK QUESTIONS
PAGES 39-46

1. lot lines
2. A. 90°
   B. 8′-0″
   C. 6′-0″
   D. 10′-0″
3. diagonals
4. D. ledger
5. A. 2′
6. B. 8″
7. D. 8000 lb.
8. A. 5″
   B. 5″
   C. 10″
   D. 20″
9. A. 12″
   B. 6″

10. D. 5/8″
11. A. Grade stake.
    B. Level.
    C. Corner stake.
    D. Outside.
12. hydration
13. D. 94 lb.
14. D. 1 1/2″
15. 4050 lb./about 2 tons
16. A. Wales.
    B. Sheathing.
    C. Tie rods/wall ties.
    D. Studs.
17. D. sill
18. A. anchor bolt
    B. 4
19. A. Face shell.
    B. Concave end.
    C. Ear.
    D. Cells or cores.
    E. Cross web.
20. B. 3/8″
21. A. Stretcher (3 core).
    B. Corner.
    C. Double-corner/pier.
    D. Bull nose.
    E. Beam/lintel.
    F. Half-cut header.
    G. Jamb.
22. See Figure 7-53 in text.
23. areaway
24. A. Tack strip.
    B. Reinforcement.
    C. Membrane dampproofing
    D. Rigid insulation.
    E. Granular fill.
25. A. 18 hours
26. C. pressure treated with chemicals
27. C. sump pit
28. blocking
29. D. 40°F
30. B. admixtures
31. 16 cu. yd.
32. 7 1/2 cu. yd.
33. 5 2/3 cu. yd.
34. 1944
35. 112.5

ANSWERS TO UNIT 7 QUIZ

1. False.
2. False.
3. True.
4. True.
5. False.
6. A. 8″
7. B. Segregation
8. B. 1800, 2500
9. C. Hydration
10. B
11. E
12. D
13. C
14. G
15. F
16. A

Reproducible Master 7-1 (Part A)
Using Batter Boards

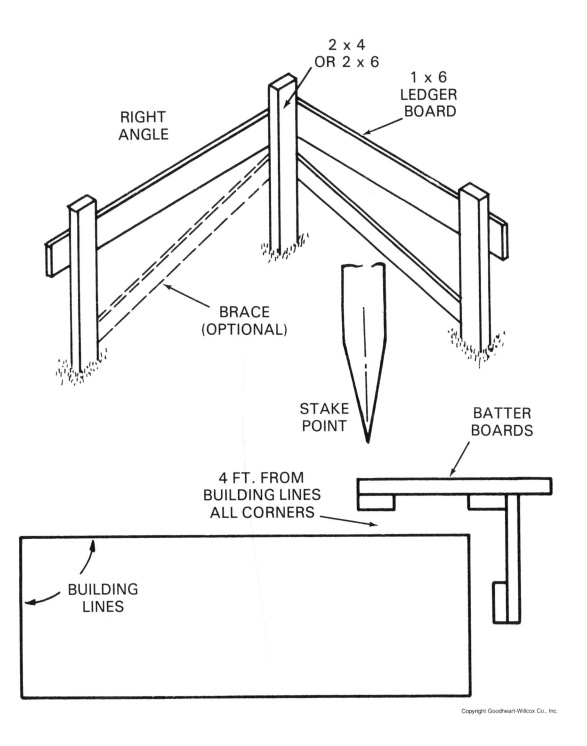

Reproducible Master 7-1 (Part B)
Using Batter Boards

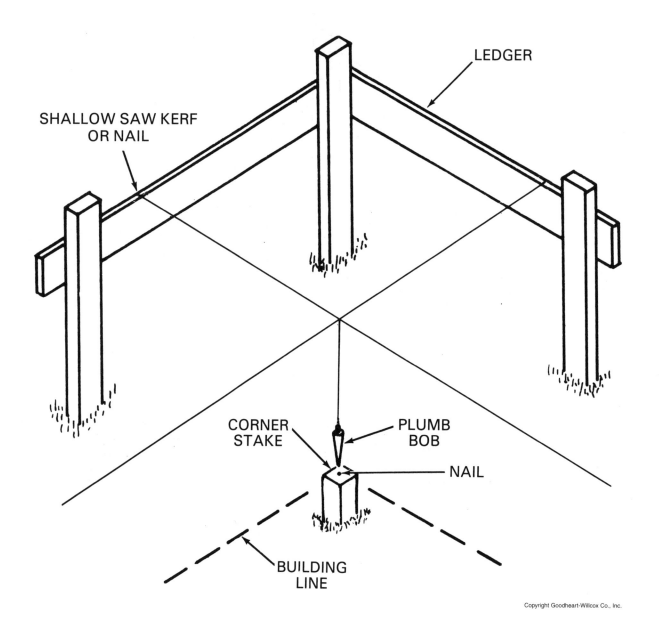

LEDGER

SHALLOW SAW KERF
OR NAIL

CORNER
STAKE

PLUMB
BOB

NAIL

BUILDING
LINE

Reproducible Master 7-2

Foundation Elements

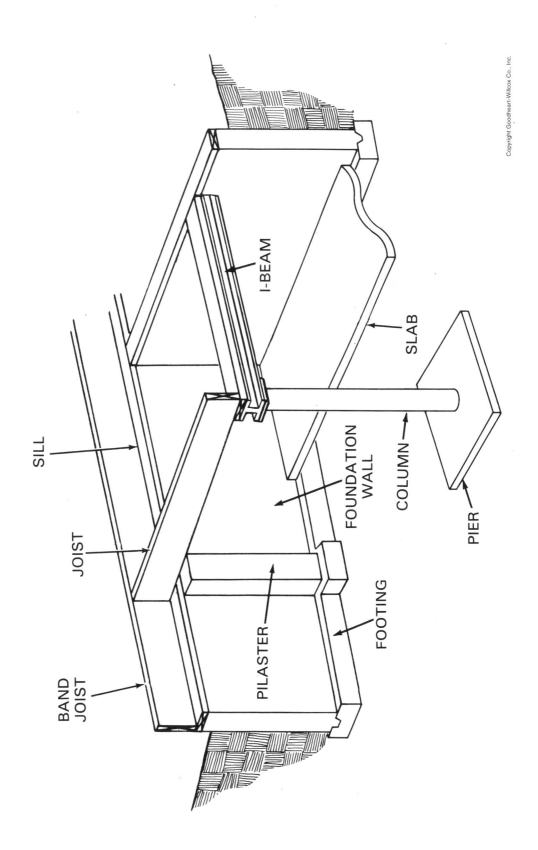

SILL

JOIST

BAND JOIST

PILASTER

FOOTING

I-BEAM

SLAB

FOUNDATION WALL

COLUMN

PIER

Foundation Types

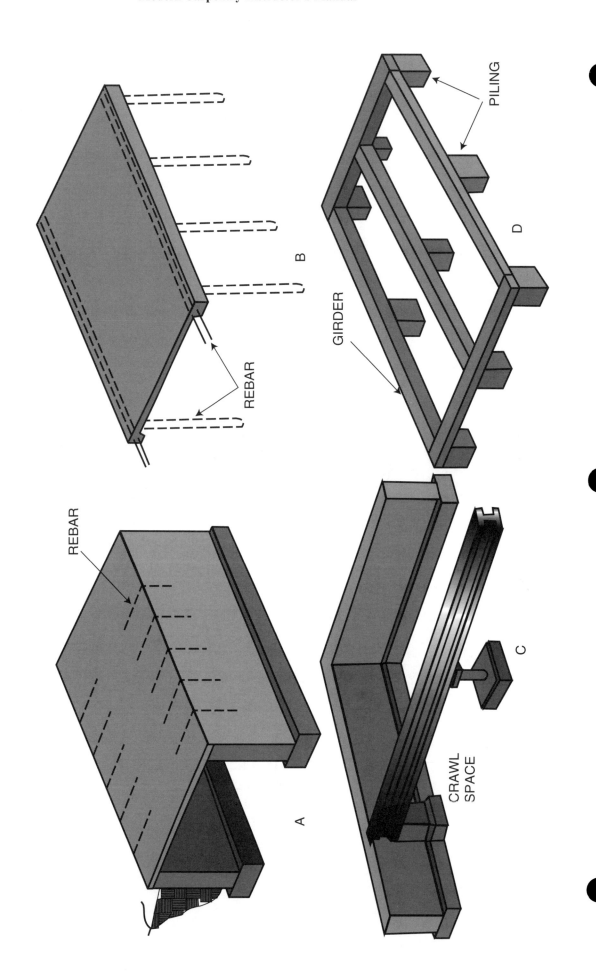

REBAR

B

GIRDER

PILING

D

REBAR

A

CRAWL
SPACE

C

Reproducible Master 7-4

Laying Out Footing Forms

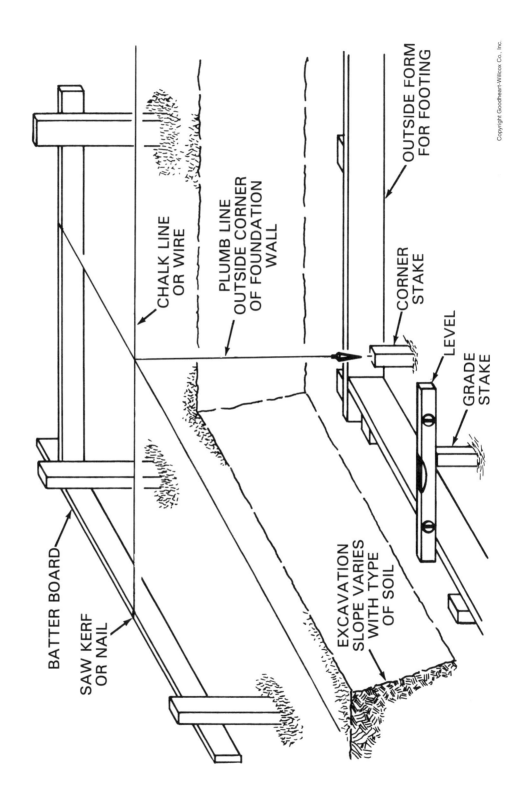

CHALK LINE OR WIRE

PLUMB LINE OUTSIDE CORNER OF FOUNDATION WALL

OUTSIDE FORM FOR FOOTING

CORNER STAKE

LEVEL

GRADE STAKE

EXCAVATION SLOPE VARIES WITH TYPE OF SOIL

BATTER BOARD

SAW KERF OR NAIL

Reproducible Master 7-5
Locating Footing Forms

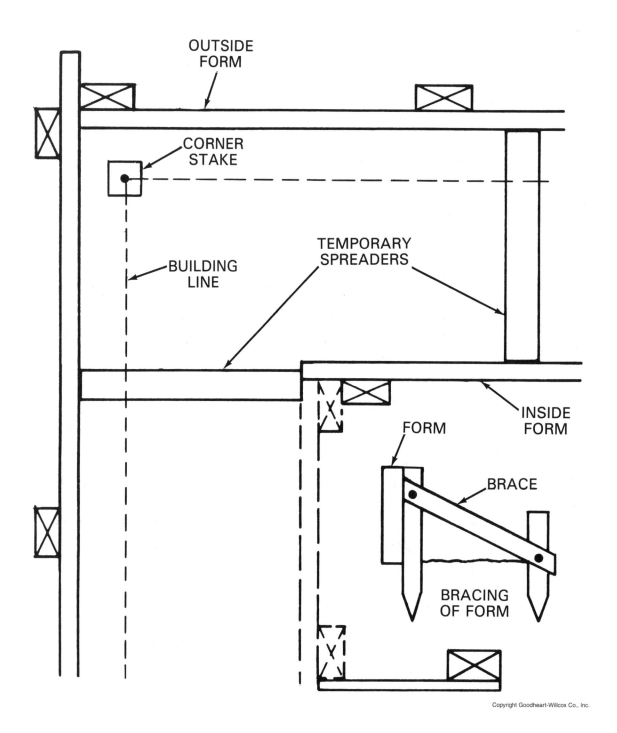

OUTSIDE
FORM

CORNER
STAKE

BUILDING
LINE

TEMPORARY
SPREADERS

INSIDE
FORM

FORM

BRACE

BRACING
OF FORM

Reproducible Master 7-6

Concrete Blocks

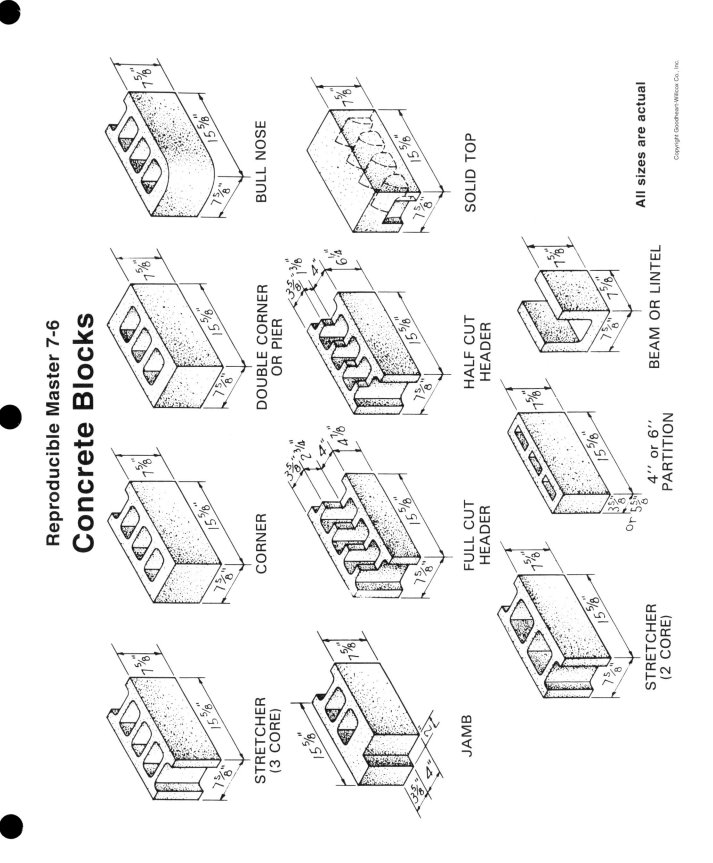

BULL NOSE

SOLID TOP

DOUBLE CORNER
OR PIER

HALF CUT
HEADER

BEAM OR LINTEL

CORNER

FULL CUT
HEADER

4″ or 6″
PARTITION

STRETCHER
(3 CORE)

JAMB

STRETCHER
(2 CORE)

All sizes are actual

Reproducible Master 7-7
Wood Foundation Design

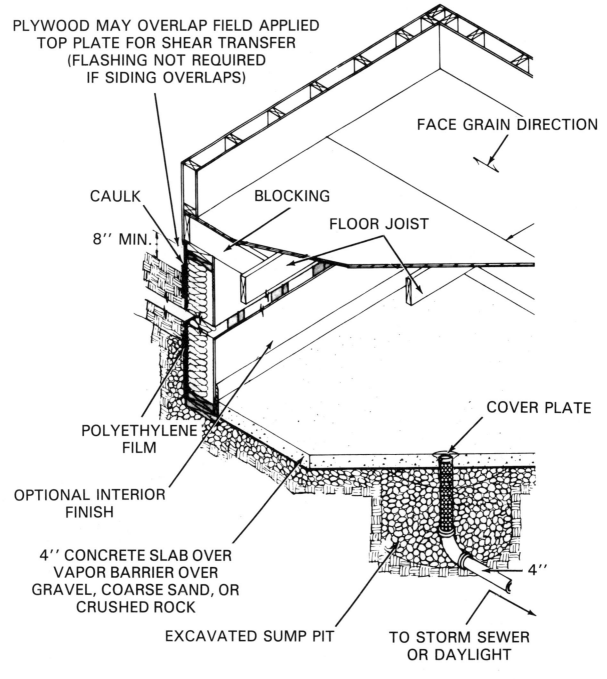

PLYWOOD MAY OVERLAP FIELD APPLIED
TOP PLATE FOR SHEAR TRANSFER
(FLASHING NOT REQUIRED
IF SIDING OVERLAPS)

FACE GRAIN DIRECTION

CAULK

BLOCKING

8″ MIN.

FLOOR JOIST

POLYETHYLENE
FILM

OPTIONAL INTERIOR
FINISH

COVER PLATE

4″ CONCRETE SLAB OVER
VAPOR BARRIER OVER
GRAVEL, COARSE SAND, OR
CRUSHED ROCK

4″

EXCAVATED SUMP PIT

TO STORM SEWER
OR DAYLIGHT

Unit 7 Procedure Checklist
Setting Up Batter Boards

Name _____ Total _____

❏ Selects proper materials for stakes, ledgers, and braces.	5	4	3	2	1	Must be told the appropriate materials to use for batter boards.
❏ Properly locates positions of stakes (approximately 4′ from intersection of building line).	5	4	3	2	1	Places stakes directly in line with building lines.
❏ Nails ledger boards in place and checks levelness; verifies levelness with other ledger boards.	5	4	3	2	1	Nails ledger boards at inconvenient height, disregarding levelness. Does not check levelness with other ledger boards.
❏ Properly runs lines and plumbs corner over corner stake; takes care in making sure alignment is accurate.	5	4	3	2	1	Cannot run nor adjust building lines; recklessly places lines.

Unit 7 Procedure Checklist
Construction Footing Forms

Name _____ Total _____

❏ Accurately re-establishes building lines.	5	4	3	2	1	Neglects to re-establish building lines.
❏ Sets up builders' level and accurately locates grade and corner stakes; connects corner stakes with lines.	5	4	3	2	1	Sets up grade and corner stakes randomly without verifying location with leveling instrument.
❏ Determines footing width and accurately lays it out.	5	4	3	2	1	Must be told the footing width; unable to determine proper placement of footing boards.
❏ Efficiently sets up outside footing boards, then the inside ones using spacers or spreaders; determines whether bracing will be required and installs it if necessary.	5	4	3	2	1	Must be assisted when setting up footing form boards; does not use spacers or spreaders; does not realize importance of bracing in necessary situations.

Unit 7 Quiz

Footings and Foundations

Name _____ Score _____

True-False

Circle T if the answer is True or F if the answer is False.

T F 1. Batter boards should be 6″ away from the corners created by the building lines.

T F 2. Reinforced footings gain their strength from aggregate and concrete.

T F 3. Monolithic concrete is constructed in one continuous pour.

T F 4. When mixing concrete, first combine the aggregate and cement before adding water.

T F 5. In most areas, concrete sidewalks are at least 12″ thick.

Multiple Choice

Choose the answer that correctly completes the statement. Write the corresponding letter in the space provided.

_____ 6. Foundations should extend about _____ above finished grade.

       A. 8″
       B. 16″
       C. 24″
       D. 32″

_____ 7. _____ is a condition in which the large aggregate gets separated from the cement paste and smaller aggregate.

       A. Separation
       B. Segregation
       C. Vibration
       D. None of the above.

_____ 8. Mortar used in concrete block foundations , according to ASTM standards, must withstand pressures from _____ to _____ psi.

       A. 1000, 1500
       B. 1800, 2500
       C. 2000, 2700
       D. 2700, 3400

_____ 9. _____ is the chemical reaction of cement and water during concrete hardening.

       A. Setting
       B. Segregation
       C. Hydration
       D. None of the above.

Identification

Identify the parts of the following foundation.

_____ 10. Slab.

_____ 11. Foundation wall.

_____ 12. Column.

_____ 13. Pier.

_____ 14. Pilaster.

_____ 15. Footing.

_____ 16. I beam.

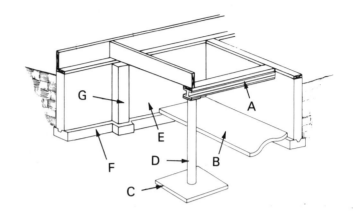

Floor Framing

OBJECTIVES

Students will be able to:

- [ ] Explain the difference between platform and balloon framing.
- [ ] Identify the main parts of a platform frame.
- [ ] Calculate the load on girders and beams used in residential construction.
- [ ] Lay out and install sills on a foundation wall.
- [ ] Describe how layouts are made on header joists.
- [ ] Explain the correct procedure to follow when assembling a wood or metal floor frame.
- [ ] Identify the parts of a floor truss.
- [ ] Prepare a sketch that shows how overhangs and projections are framed.
- [ ] Describe kinds of material used for subflooring.
- [ ] Estimate materials (sizes and amounts) required to construct a specific floor frame.

INSTRUCTIONAL MATERIALS

Text: Pages 179-204

Important Terms, page 202
Test Your Knowledge, page 202-203
Outside Assignments, page 203

Workbook: Pages 47-54

Instructor's Manual:

Reproducible Master 8-1, *Platform Framing Details*
Reproducible Master 8-2, *Framing Methods at Sill*
Reproducible Master 8-3, *Using Joist Hangers*
Reproducible Master 8-4, *Framing over Girders/Beams, Parts A, B,* and *C*
Reproducible Master 8-5, *Laying Out Joists, Parts A* and *B*
Reproducible Master 8-6, *Floor Frame*
Reproducible Master 8-7, *Assembling Floor Frame, Parts A* and *B*
Reproducible Master 8-8, *Truss Construction*

Reproducible Master 8-9, *Placing Subflooring*
Procedure Checklist—*Installing Sills*
Procedure Checklist—*Installing Joists*
Unit 8 Quiz

TRADE-RELATED MATH

A carpenter may be required to estimate the number of floor joists used for a job. First, use a plan to scale the length of joists needed. Be sure to allow additional length so the joists rest properly on girders and partitions. Then, multiply the length of the wall by 3/4 for 16″ O.C. (3/5 for 20″ O.C.; 1/2 for 24″ O.C.) and add 1. Be sure to add extra pieces for doubled joists. The formula looks like this:

$$\text{length of wall} \times \frac{3}{4} + 1 + \text{extras} = \text{Number of joists}$$

For example, if the wall length is 28′, the joists are spaced 16″ O.C., and 4 extra joists are required, the number of joists is figured as follows:

$$\text{length of wall} \times \frac{3+1}{4} + \text{extras} = \text{Number of joists}$$

$$28 \times \frac{3}{4} + 1 + 4 = 21 + 1 + 4 = 26$$

26 joists will be required

INSTRUCTIONAL CONCEPTS AND STUDENT LEARNING EXPERIENCES

TYPES OF FRAMING

1. Using Figure 8-1, discuss the common types of framing for residential structures. Have the students identify the type of framed structures they live in.
2. Discuss the criteria that are considered when determining the type of framing used in a particular location.
3. Explain the two basic types of framing. Describe the difference between the two types.

4. Discuss the applications for platform framing. Discuss why platform framing has become the prevalent type of framing method.
5. Using Reproducible Master 8-1, *Platform Framing Details*, discuss the construction procedures used for platform framing.
6. Using Reproducible Master 8-2, *Framing Methods at Sill*, discuss the framing methods shown in an architectural plan. Discuss what part of the plans contain details showing the framing. Have the students note the differences
7. Discuss modern applications for balloon framing. Explain why balloon framing is now used on a limited basis.

GIRDERS AND BEAMS

1. Discuss the reasons for using girders and beams in a residential structure. Have students identify the types of materials commonly used as girders or beams.
2. Describe the procedure used to determine the size of girder or beam needed for support in a structure.
3. List and discuss the advantages and disadvantages of using steel beams in a house.
4. Explain the difference between S and W beams.
5. Discuss the procedure for constructing a built-up girder. Stress that the lumber should be securely nailed together, and that the joints should rest directly over columns or posts.

POSTS AND COLUMNS

1. Discuss the sizes of beams that are generally used to support beams or girders. Emphasize that, as a rule of thumb, it is safe to assume that a post or column whose greatest dimension is equal to the width of the girder it supports will adequately carry the imposed loads.
2. Describe the types of footings that must be provided to support girder posts and columns. Stress that some forethought must be used to make sure that reinforcing rod or iron bolts are embedded into the concrete footing before the concrete sets.
3. Have students identify the types of materials used for posts and columns. If possible, set up a steel adjustable post and demonstrate its use.

FRAMING OVER GIRDERS AND BEAMS

1. Show how joist hangers are placed using Reproducible Master 8-3, *Using Joist Hangers*. Discuss the advantages of using joist hangers.

2. Demonstrate how joist hangers are attached to a girder or beam.
3. Using parts *A, B,* and *C,* of Reproducible Master 8-4, *Framing over Girders/Beams,* discuss the common methods of framing joists over girders and beams. Discuss the means for raising or lowering the ceiling height by using different methods of framing joists.

SILL CONSTRUCTION

1. Discuss the purpose of a sill. Have students list other names that are commonly used for a sill.
2. Describe why 2×6 lumber is typically used for sills. Explain that in the past, 2×4 lumber was commonly used for sills.
3. Ask the students to describe any type of preparation (such as a sill sealer) that should be used prior to attaching the sill.

TERMITE SHIELDS

1. Have students identify the parts of the United States that are susceptible to termites. Discuss the destruction that may be caused by termites.
2. Describe techniques that may be used to guard against termite damage including termite shields, treated lumber, and poisoning the soil around the structure.

INSTALLING SILLS

1. Discuss the importance of accurately laying out the holes in a sill when it is to be attached with anchor bolts. Using part *A* of Reproducible Master 8-5, *Laying Out Joists,* describe the procedures that can be used to attach a sill to a foundation wall. Stress that sill will be set back from the outside of the foundation wall, and that this measurement must be taken into consideration when laying out the anchor bolt holes.
2. Using the scale model of the foundation wall constructed in Unit 6, have the students attach a sill to it. Be sure to include sill sealer.
3. Describe what should be done when an uneven foundation wall is encountered.

JOISTS

1. Discuss the purpose of floor joists. Explain that joists are usually placed 16″ O.C., but that other dimensions may be used in some circumstances.
2. Using a complete set of plans for a residential structure, have the students locate information that should be used to determine the type and direction of the floor joists. Have the students take

note of dimensions and other notes that relate to other superstructure members.

3. Using part *B* of Reproducible Master 8-5, *Laying Out Joists,* discuss where joist positions are usually located. Mention that in platform framing, joist position is generally marked on the joist header, not on the sill. Stress the importance of marking an X on the sill or joist header to indicate which side of the line that the joist should actually be located.

4. Have the students list areas where they feel that joists should be doubled, such as under load-bearing partitions, and around openings in the floor frame for stairways, chimneys, and fireplaces.

5. Discuss the proper procedure for installing joists. Emphasize that joists with a crown should be placed with the crown facing up.

6. Using Reproducible Master 8-6, *Floor Frame,* point out the framing members that must be installed in a finished floor assembly.

7. Explain that header and tail joist cuts must be made square, and that they must fit tightly together with other framing members. Stress that considerable strength will be lost if members do not fit tightly together.

8. Using Figure 8-33, describe the procedure for assembling frames for floor openings.

9. Using parts *A* and *B* of Reproducible Master 8-7, *Assembling Floor Frame,* describe a good nailing pattern to use when assembling a floor frame. Emphasize that a good nailing pattern will provide good support for concentrated loads.

10. Discuss the purpose of bridging. Cite examples of places where bridging can be eliminated.

11. Describe the various types of bridging used for residential structures including wood cross bridging, solid bridging, and prefabricated steel bridging. If possible, obtain samples of different types of bridging and allow students to inspect each type.

12. Demonstrate how wood cross bridging is laid out using a framing square. Also have students demonstrate the proper layout procedure.

SPECIAL FRAMING PROBLEMS

1. Describe special framing problems that may be encountered in modern residential construction. Using Figure 8-41, discuss the problems that may be encountered when floor joists run parallel to the foundation wall.

2. Identify other areas of a house that may pose special framing problems, such as bathrooms and

areas where tile or stone is to be laid.

3. Demonstrate the concepts of tension and compression in regard to floor joists. Secure a thin piece of stock in two vices (one at each end) along the edge of a shop table. Hang weights along the bottom of the stock to represent the loads of a floor, furniture, and appliances. Note the deflection of the stock. Drill a hole along the top edge of the strip, hang the weights, and note the deflection. Make a cut three-quarters of the way through a similar piece of stock. Hang weights from the bottom (as done in the previous examples) and note the deflection and/or damage. Use this demonstration to stress the importance of properly placing cuts or drilling holes in joists.

4. Discuss precautions that should be considered when cutting joists for plumbing runs.

5. Explain the advantages of using a crawl space in a residential structure. Discuss the construction of a crawl space that can be used as a heating/cooling plenum.

FLOOR TRUSSES

1. Discuss the purpose of floor trusses in residential construction. Cite advantages of using floor trusses.

2. Using Reproducible Master 8-8, *Truss Construction,* identify the various types of truss construction.

3. Discuss the purpose of a subfloor. Have students identify the types of materials commonly used for subfloors.

4. Describe the advantages of using plywood as a subflooring.

5. Using Reproducible Master 8-9, *Placing Subflooring,* discuss the proper spacing and nailing procedures for plywood subflooring.

OTHER SHEET MATERIALS

1. Have students identify other sheet materials that can be used as subflooring.

GLUED FLOOR SYSTEM

1. Discuss the advantages of using a glued floor-system. Describe the correct procedure for applying glue for subflooring.

STEEL JOISTS

1. Identify applications for using steel joists in residential construction.

2. Discuss methods used to fasten steel joists to wood headers and sills. Also discuss how subflooring is fastened to steel joists.

ESTIMATING MATERIALS

1. Demonstrate the method used to estimate the number and size of floor joists for a residential structure.
2. Discuss the means for estimating the amount of subflooring required.

UNIT REVIEW

1. Review the unit objectives. Be sure that the students fully understand each objective.
2. Assign *Important Terms, Test Your Knowledge* questions, and *Outside Assignments* on pages 202 and 203 of the text. Review the answers in class.
3. Assign pages 47-54 of the workbook. Review the answers in class.

EVALUATION

1. Use Procedure Checklist—*Installing Sills,* to evaluate the techniques that students used to perform these tasks.
2. Use Procedure Checklist—*Installing Joists,* to evaluate the techniques that students used to perform these tasks.
3. Use Unit 8 Quiz for in-class evaluation. Correct the quizzes and return them to the students for review.

ANSWERS TO TEST YOUR KNOWLEDGE TEXT PAGE 202

1. platform
2. hangers
3. sheathing thickness
4. sill
5. S and W
6. 1/360
7. tail
8. trimmer
9. two
10. center
11. 3
12. parallel
13. Driving a steel pin through a flange on the header into the wall.
14. False.

ANSWERS TO WORKBOOK QUESTIONS PAGES 47-54

1. backfill
2. A. 1 1/2 story.
   B. Two-story.
   C. Split level.
3. A. Joist header.
   B. Subfloor.
   C. Joist.
   D. Sole plate.
   E. Sill.
   F. Girder.
4. western
5. C. ribbon
6. C. 40
7. B. 12″
8. A. Girder.
   B. Ledger.
   C. Joist.
   D. Solid bridging/blocking.
   E. Steel beam.
   F. Steel post.
9. C. attach sill to foundation walls
10. A. Width.
    B. Depth.
    C. Flange.
    D. Web.
11. 1/3″
12. A. 16″
    B. 16″
    C. 15 1/4″
13. B. 3-16d
14. C. one-half
15. A. Header.
    B. Regular joist.
    C. Tail joist.
    D. Double header.
    E. Double trimmer.
    F. Tail joist.
16. D. 12′-0″.
17. A. Make each joist stronger.
18. A. 7′-0″
    B. 4′-8″
19. doubled
20. B. approximately in the middle
21. B. 1/2″
22. A. 25%
23. See Figure 8-44 in text.
24. A. 12″
    B. 6″
25. A and B
26. Wear leather gloves to avoid cuts.
    Let welded joints cool before handling them.

27. Headers 4: 2 × 8 × 16.
    Joists 25: 2 × 8 × 12.
    Total bd. ft.: 486.
28. Shiplap: 507 bd. ft.
                384 sq. ft.
    Plywood: 12 pcs.
                384 sq. ft.
29. 204′
    204 bd. ft.
30. 114 pcs.
    2432 bd. ft.
31. 288 lineal ft.
    72 bd. ft.
32. 48 pcs.
    1536 sq. ft.
33. Use smaller joists, doubling them up or spacing them closer together.

ANSWERS TO UNIT 8 QUIZ

1. True.
2. True.
3. False.
4. True.
5. C. S and W
6. B. engineered
7. D. All of the above.
8. A. trimmers
9. C
10. B
11. A
12. G
13. E
14. D
15. F

Modern Carpentry Instructor's Manual

Reproducible Master 8-1

Platform Framing Details

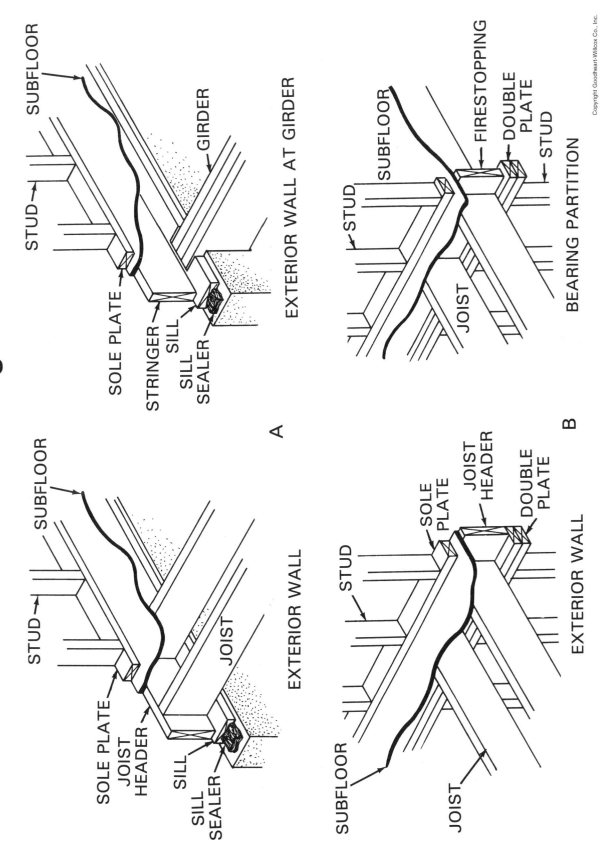

STUD

SUBFLOOR

SOLE PLATE

STRINGER

SILL

SILL SEALER

GIRDER

EXTERIOR WALL AT GIRDER

A

STUD

SUBFLOOR

SOLE PLATE

JOIST HEADER

SILL

SILL SEALER

JOIST

EXTERIOR WALL

STUD

SUBFLOOR

FIRESTOPPING

DOUBLE PLATE

STUD

JOIST

BEARING PARTITION

B

STUD

SOLE PLATE

JOIST HEADER

DOUBLE PLATE

SUBFLOOR

JOIST

EXTERIOR WALL

Reproducible Master 8-2
Framing Methods at Sill

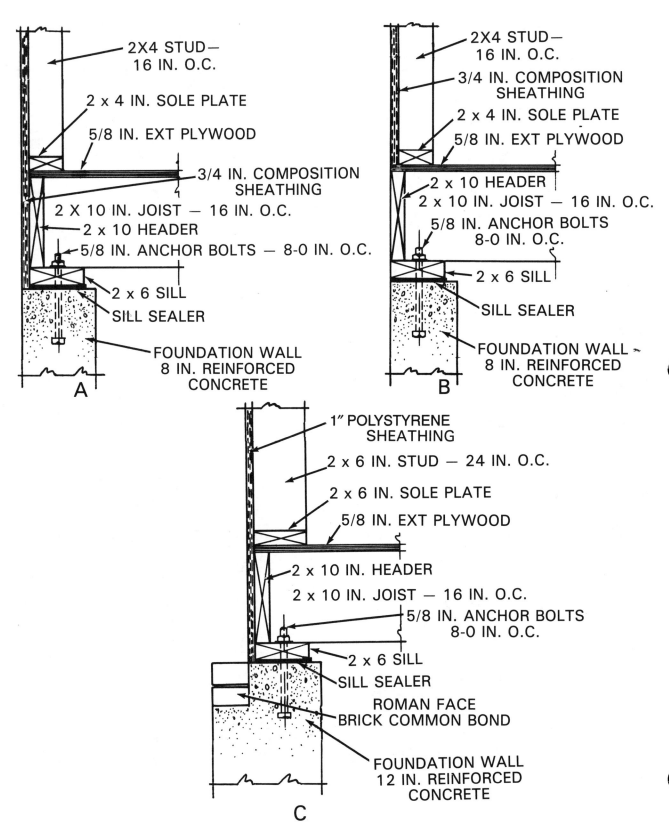

Reproducible Master 8-3
Using Joist Hangers

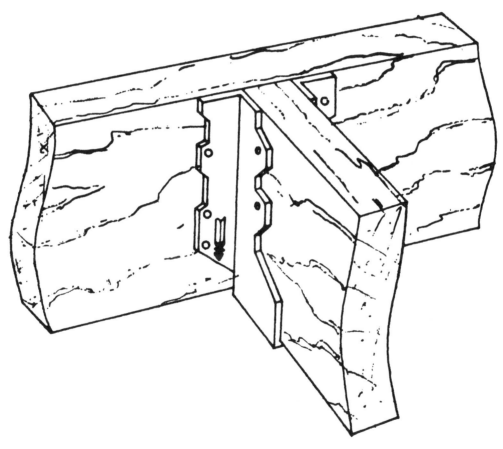

Reproducible Master 8-4 (Part A)
Framing over Girders/Beams

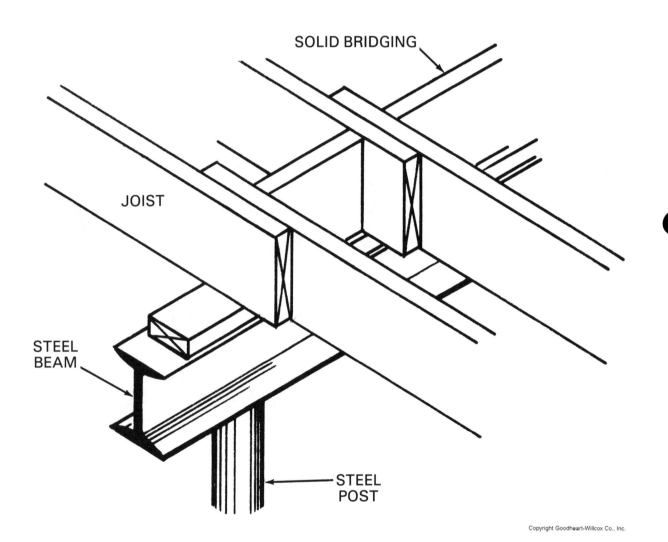

SOLID BRIDGING

JOIST

STEEL
BEAM

STEEL
POST

Reproducible Master 8-4 (Part B)
Framing over Girders/Beams

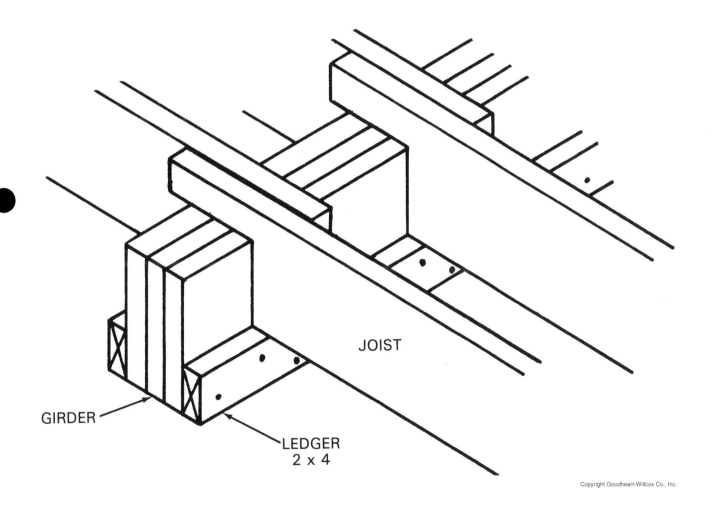

JOIST

GIRDER

LEDGER
2 x 4

Reproducible Master 8-4 (Part C)
Framing over Girders/Beams

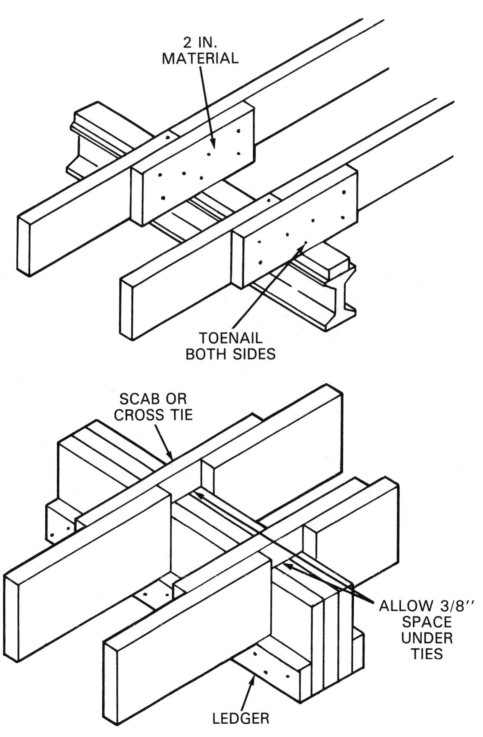

2 IN.
MATERIAL

TOENAIL
BOTH SIDES

SCAB OR
CROSS TIE

ALLOW 3/8''
SPACE
UNDER
TIES

LEDGER

Reproducible Master 8-5 (Part A)
Laying Out Joists

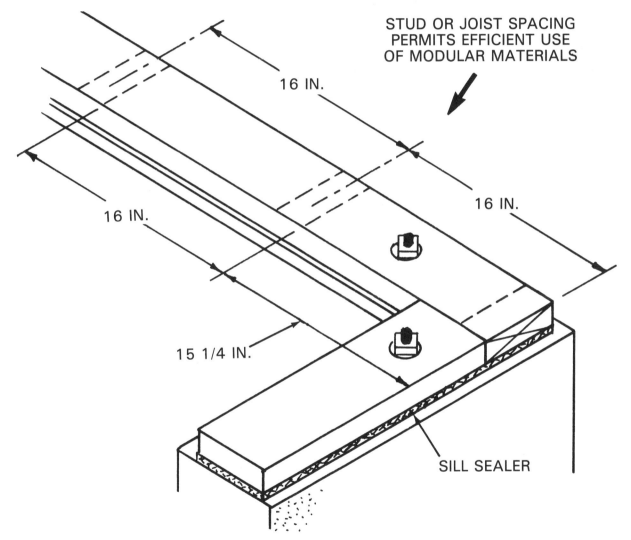

STUD OR JOIST SPACING
PERMITS EFFICIENT USE
OF MODULAR MATERIALS

16 IN.

16 IN.

16 IN.

15 1/4 IN.

SILL SEALER

Reproducible Master 8-5 (Part B)
Laying Out Joists

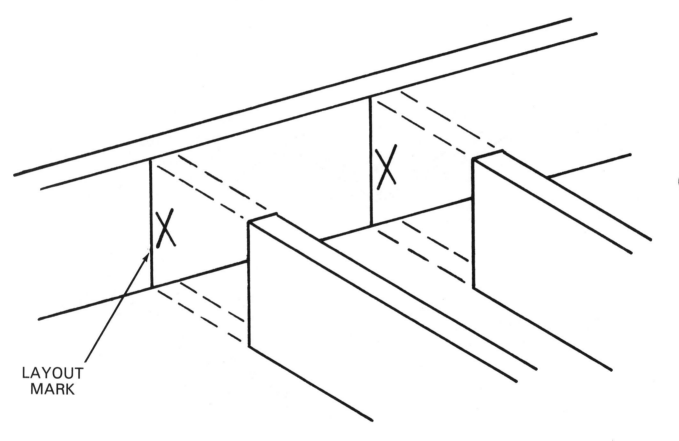

LAYOUT
MARK

Reproducible Master 8-6

Floor Frame

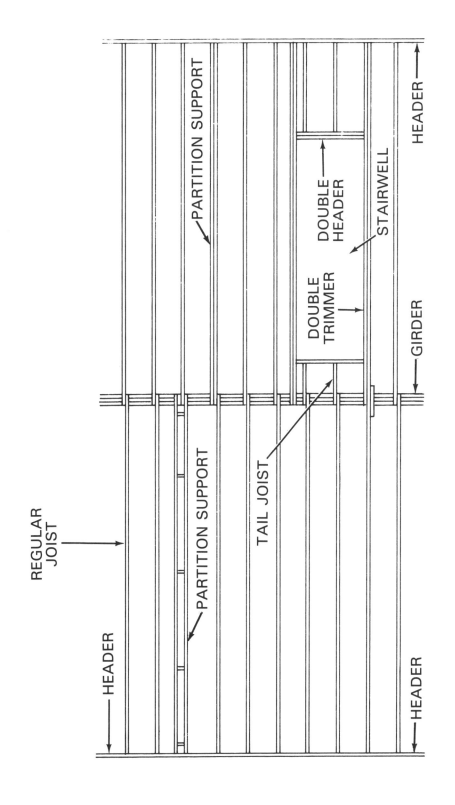

119

Reproducible Master 8-7 (Part A)
Assembling Floor Frame

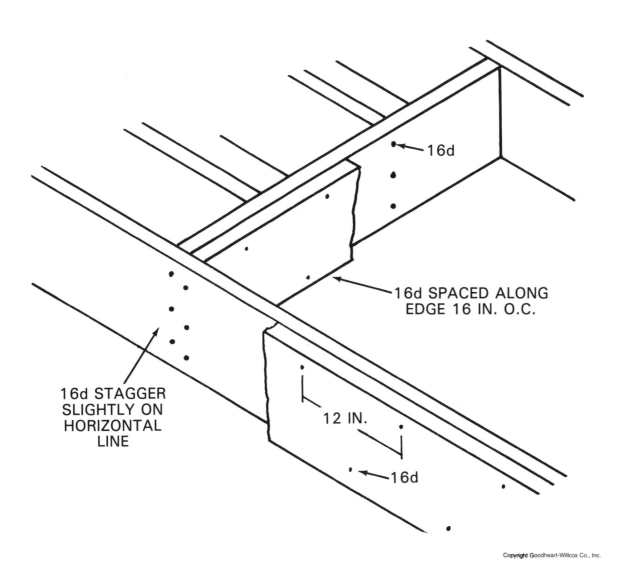

16d

16d SPACED ALONG
EDGE 16 IN. O.C.

16d STAGGER
SLIGHTLY ON
HORIZONTAL
LINE

12 IN.

16d

Reproducible Master 8-7 (Part B)
Assembling Floor Frame

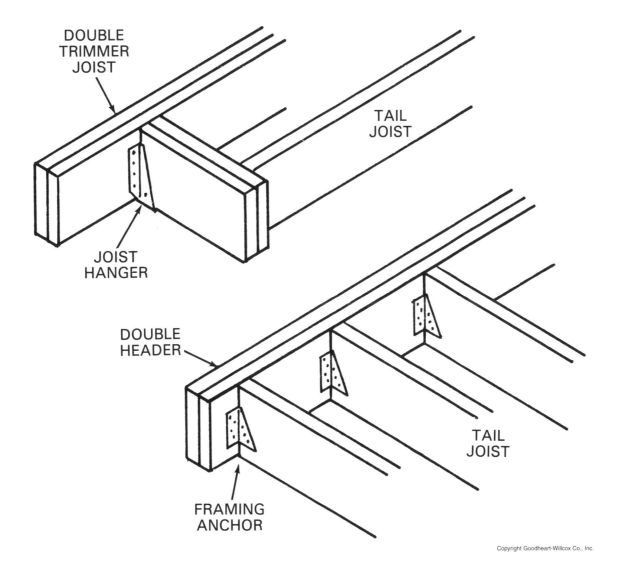

DOUBLE
TRIMMER
JOIST

TAIL
JOIST

JOIST
HANGER

DOUBLE
HEADER

TAIL
JOIST

FRAMING
ANCHOR

Reproducible Master 8-8

Truss Construction

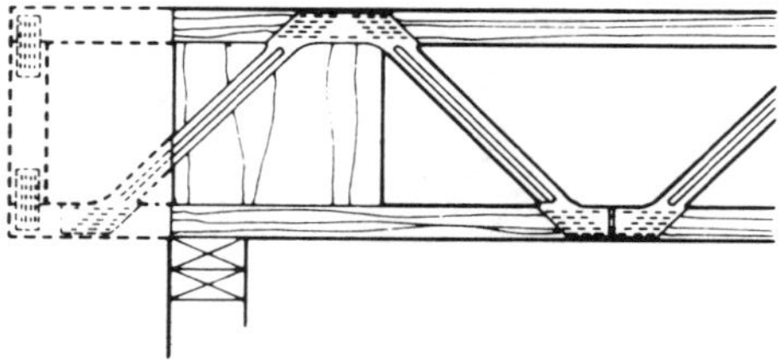

BOTTOM CHORD CANTILEVER —ON PANEL

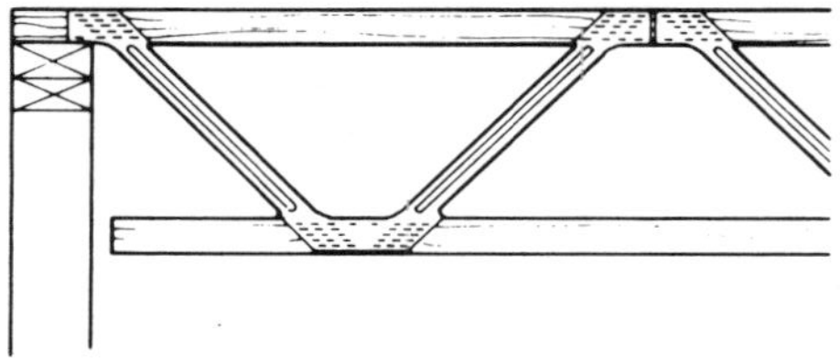

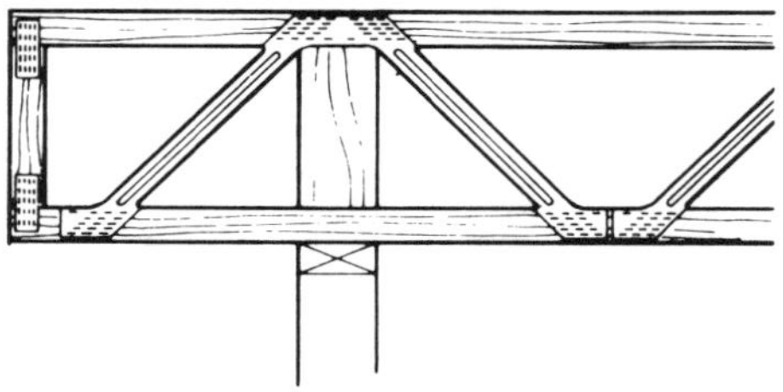

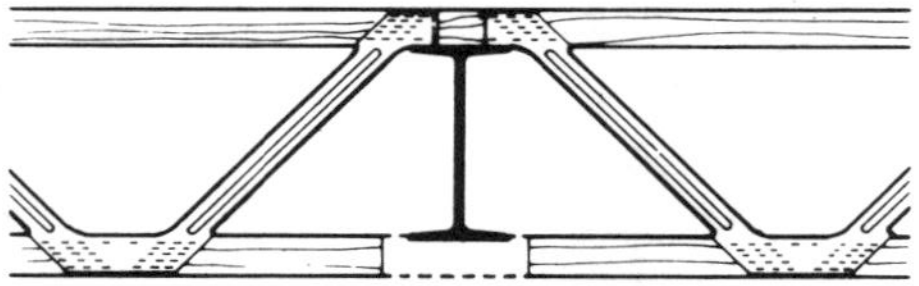

BOTTOM CHORD WITH CONTINUOUS BANDING

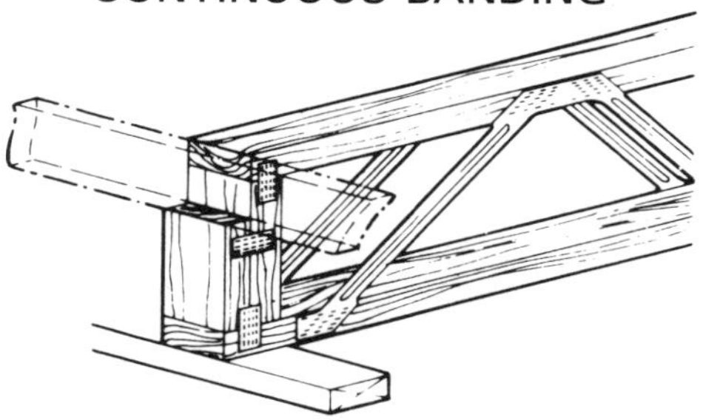

FIELD CUT TRUSS DETAIL

CENTER BEAM

TOP CHORD

BOTTOM CHORD W/O BANDING BLOCK

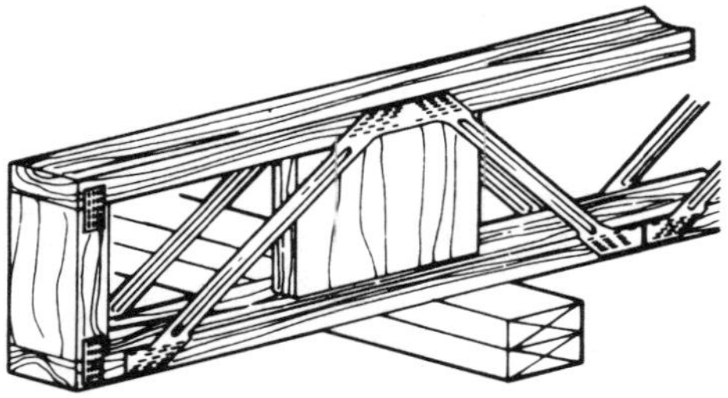

BALCONY JOIST DETAIL

1/2'' SPACE BLOCK

1/2'' SPACER BLOCK

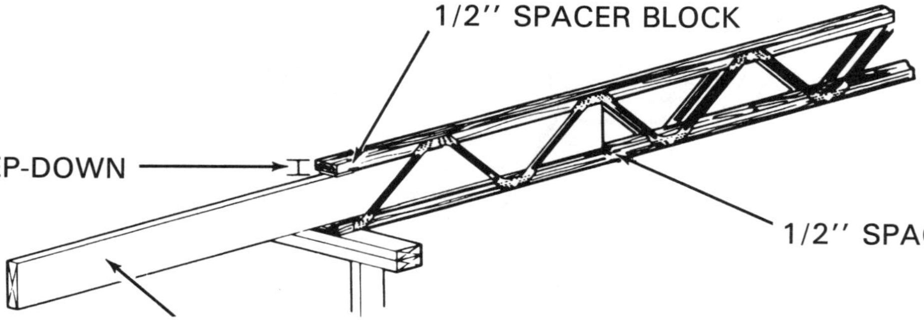

1/2'' SPACER BLOCK

2'' STEP-DOWN

2 x 8 BALCONY JOIST

Placing Subflooring

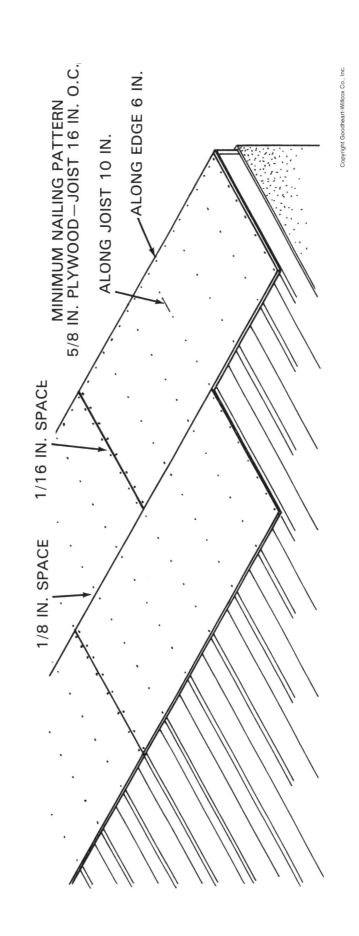

MINIMUM NAILING PATTERN
5/8 IN. PLYWOOD—JOIST 16 IN. O.C.

ALONG JOIST 10 IN.

ALONG EDGE 6 IN.

1/16 IN. SPACE

1/8 IN. SPACE

Unit 8 Procedure Checklist
Installing Sills

Name _____ Total _____

❏ Removes nuts and washers; lays sill along foundation wall for anchor bolt placement.	5	4	3	2	1	Tries to measure position of anchor bolt holes.
❏ Lays out lines across sill for anchor bolt placement accurately measures distance from center of the bolt to the outside of foundation wall and then subtracts sheathing thickness; accurately lays out this dimension on the sill.	5	4	3	2	1	Guesses at placement of anchor bolt holes; forgets to compensate for sheathing thickness.
❏ Snaps line on foundation wall to accurately align sill.	5	4	3	2	1	Neglects to snap line on foundation wall.
❏ Bores anchor bolt holes at one time for maximum productivity, using the proper size holes (approximately 1/4″ larger than bolt diameter).	5	4	3	2	1	Bores anchor bolt hole as each is laid out, thus reducing productivity; bores holes that are the same size as the bolts.
❏ Dry fits sills on foundation wall; removes sills; installs sill sealer and replaces sill; tightens sills down using washers and nuts.	5	4	3	2	1	Places sills on foundation wall without sill sealer; neglects to usc washers when tightening sills down.
❏ Verifies proper placement of sill on foundation wall; levels sill and shims where necessary.	5	4	3	2	1	Neglects to check placement of sill; forgets to level the sill.

Unit 8 Procedure Checklist
Installing Joists

Name _____ Total _____

❑ Toenails header joist to the sill.	5	4	3	2	1	Tries to balance the header joist on the sill.
❑ Positions full length joists with crowns turned up; nails joists into position using three 16d nails.	5	4	3	2	1	Neglects to check crown on joists; uses fewer or more nails than necessary.
❑ Fastens joists along the opposite wall into position; properly joins joists that butt or overlap.	5	4	3	2	1	Neglects to fasten joists along opposite wall into position; does not join butting or overlapping joists.
❑ Interprets plans; accurately locates areas where doubled joists are required.	5	4	3	2	1	Neglects to install doubled joists.

Unit 8 Quiz
Floor Framing

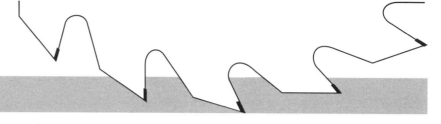

Name _____ Score _____

True-False

Circle T if the answer is True or F if the answer is False.

T   F   1. Platform framing is also referred to as western framing.

T   F   2. In balloon framing, studs extend from the sill to the rafter plate.

T   F   3. When cutting openings in joists, cuts should be made along the bottom edge.

T   F   4. Floor joists carry the weight of the floor between the sills and girders.

Multiple Choice

Choose the answer that correctly completes the statement. Write the corresponding letter in the space provided.

_____   5. The two types of steel beams commonly used in residential construction are _____ beams.
  A. I and S
  B. I and W
  C. S and W
  D. None of the above.

_____   6. Units such as truss joists, wood I beams, open web joists, and steel joists, are know collectively as _____ lumber.
  A. manufactured
  B. engineered
  C. laminated
  D. span-rated

_____   7. The subfloor _____.
  A. adds rigidity to the structure
  B. provides a base for finish flooring materials
  C. provides a surface for the carpenter to lay out and construct framing
  D. All of the above.

_____   8. Joists that are doubled around openings in the floor frame are called _____.
  A. trimmers
  B. headers
  C. footers
  D. sills

Identification

Identify the parts of the platform framing detail.

_____ 9. Joist.

_____ 10. Subfloor.

_____ 11. Stud.

_____ 12. Sole plate.

_____ 13. Sill.

_____ 14. Sill sealer.

_____ 15. Joist header.

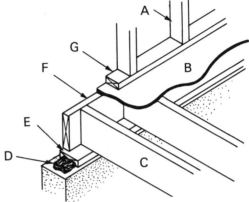

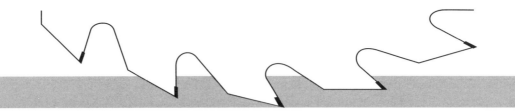

9

Wall and Ceiling Framing

OBJECTIVES

Students will be able to:
- ❏ Identify the main parts of a wall frame.
- ❏ Explain methods of forming the outside corners and partition intersections of wall frames.
- ❏ Show how rough openings are handled in wall construction.
- ❏ Explain plate and stud layout.
- ❏ Describe the construction and erection of wall sections and partitions.
- ❏ List the materials commonly used for sheathing.
- ❏ Demonstrate the process of ceiling frame construction.
- ❏ Explain and demonstrate construction of walls using metal studs.
- ❏ Estimate materials required for wall frames, ceiling frames, and sheathing.

INSTRUCTIONAL MATERIALS

Text: Pages 205-232
    Important Terms, page 231
    Test Your Knowledge, page 231
    Outside Assignments, page 231
Workbook: Pages 55-62
Instructor's Manual:
    Reproducible Master 9-1, *Framing Corners*
    Reproducible Master 9-2, *Framing Wall Intersections*
    Reproducible Master 9-3, *Framing Wall Openings*
    Reproducible Master 9-4, *Framing with Patterns, Parts A, B,* and *C*
    Procedure Checklist—*Constructing Wall Sections*
    Procedure Checklist—*Erecting Wall Sections*
    Unit 9 Quiz

TRADE-RELATED MATH

Sheathing adds rigidity, strength, and some insulating qualities to the wall. To determine the amount of sheathing, first measure the perimeter of the structure. Multiply this by the wall height when sheathing extends over the foundation wall. Then, subtract the square footage of the openings in the wall to determine the area of the walls. Divide this number by the area per sheet of sheathing. The following formula is used:

$$\frac{\text{perimeter height} \times \text{wall openings}}{\text{area of one sheet of sheathing}} = \begin{array}{l}\text{Number of}\\\text{sheathing}\\\text{panels}\end{array}$$

Assume that 4 × 8 sheets of sheathing are used. The perimeter of the structure measures 20′ × 25′, with window and door openings amounting to 105 square feet. The wall height is 8′. The number of sheets of sheathing that are required is calculated as follows (round the answer to the next highest whole number):

$$\frac{\text{perimeter height} \times \text{wall openings}}{\text{area of one sheet of sheathing}} = \begin{array}{l}\text{Number of}\\\text{sheathing}\\\text{panels}\end{array}$$

$$\frac{(20 \times 25) \times 8 - 105}{32} = \frac{3895}{32} = 121.7$$

122 sheets of sheathing will be required

INSTRUCTIONAL CONCEPTS AND STUDENT LEARNING EXPERIENCES

PARTS OF THE WALL FRAME
1. Refer to Figures 9-2 and 9-3. Identify the parts of a typical wall frame including the sole plates, top plates, studs, headers, and sheathing. List the type

of materials that are used for each of these parts. Have students note where additional studs are used.

2. Using Reproducible Master 9-1, *Framing Corners,* discuss the various methods used to frame wall corners.

3. Demonstrate the methods used to frame corners by building scale models. Follow up the demonstration by having the students construct scale models of corner framing.

3. Using Reproducible Master 9-2, *Framing Wall Intersections,* discuss the methods used to construct partition intersections. Explain the importance of creating a square inside corner to provide a nailing surface for wall-covering materials.

4. Review the house plan included in Unit 6, or a comparable house plan, to determine the location of rough openings in a wall. Stress that rough opening measurements are usually made from the corners to the centerlines of the openings.

5. Discuss the purpose of headers in wall construction. Have students identify places where headers will be installed.

6. Discuss the construction of window and door headers. Show the students how to determine the length of the header by adding the rough opening plus two headers. Demonstrate the construction of a header.

7. Using Reproducible Master 9-3, *Framing Wall Openings,* review the parts of a wall frame. Stress the importance of using the proper terminology when referring to the parts of the wall frame.

8. Discuss alternate methods of header construction. Caution students about the disadvantages of using a header that extends completely to the top plate.

PLATE LAYOUT

1. Discuss different methods of laying out the plates. Stress that they should be laid out along the outside wall to minimize the necessity of moving the completed wall around the structure.

2. Refer to part *A* of Reproducible Master 9-4, *Framing with Patterns.* Describe the meaning of the Xs, Ts, and Cs on the plates. Emphasize that markings made during layout will help to eliminate confusion when constructing the wall plate.

3. Using a set of drawings from a small residential structure, have a group of students interpret the drawings and lay out the sole and top plates. Have another group of students check the accuracy of their measurements.

STORY POLE

1. Describe the purpose of a story pole. Discuss the advantages of using a story pole over laying out all the measurements individually. Stress the importance of making sure that the measurements on the story pole are accurate.

2. Using part *B* of Reproducible Master 9-4, *Framing with Patterns,* show how the story pole is used. Lay out a story pole using measurements from the plans used in the previous step.

MASTER STUD LAYOUT

1. Compare the story pole in the previous step to the layout of the master stud in part *C* of Reproducible Master 9-4, *Framing with Patterns.*

CONSTRUCTING WALL SECTIONS

1. Discuss the advantages of using P.E.T. lumber in constructing a wall section.

2. Have students list the steps involved in the construction of a common wall section. Use the listed steps to construct a wall section. Use the measurements from previously used plans.

3. Review the safety procedures that should be followed when constructing wall sections using a power nailer.

4. Attach the sheathing to the wall section. Make certain that the wall frame is square before attaching it. Mention that the same procedure for checking the squareness of a building site (measuring diagonals) can be used to check the squareness of a wall section.

ERECTING WALL SECTIONS

1. Discuss different methods that can be used to erect a wall section. If wall jacks are to be used, demonstrate their use before raising the wall section.

2. Stress the importance of plumbing the wall section after erecting it. Demonstrate the use of a straightedge to check the plumbness if the framing member or surface is warped.

PARTITIONS

1. Discuss the next stage of construction—erecting bearing partitions. Using the set of plans from the previous steps, have students determine which partitions are bearing partitions, and their location. Review the use of a chalk line at this point.

2. Stress the importance of getting the structure closed in before erecting nonbearing partitions. Discuss possible scenarios that may occur if the structure does not get closed in quickly.

3. Identify special framing considerations that may need to be addressed when constructing walls and partitions. Include openings for the heating ducts and extra bracing for the bathtub and wall-mounted stool.
4. Discuss special construction that is required for electrical runs and plumbing and venting pipes.
5. Referring to Figure 9-24, identify special reinforcement methods for studs when notched for plumbing or wiring.
6. Have students refer to the local building code to determine the wall bracing requirements.
7. Identify various types of wall bracing that can be used, including wood let-in braces and metal strap bracing. Stress that bracing *cannot* simply be attached to the outside of the wall sections because it interferes with sheathing.

DOUBLE PLATE
1. Describe why a double top plate is used.

SPECIAL FRAMING
1. Discuss special framing problems that may be encountered in residential construction.
2. Using a set of plans for a small residential structure, have students identify special framing problems that may be encountered. Have the students sketch possible framing configurations for each of the problems they identify.

WALL SHEATHING
1. Identify materials that are commonly used for wall sheathing. Discuss the advantages or disadvantages of installing wall sheathing after the wall sections have been erected.
2. Describe how fiberboard, plywood, and insulating panels are attached to the wall section. Have students explain additional factors that should be considered such as use of plywood at corners as bracing (or letting in of angle braces at corners) when fiberboard or insulating board is used as sheathing.

CEILING FRAME
1. Discuss the similarities between ceiling framing and floor framing. Identify the factors that should be considered when determining the length of span and spacing used for ceiling joists. Mention that this information is usually found in the set of plans.
2. Refer to Figure 9-49 and discuss the construction of the ceiling frame. If time and materials allow,

construct a scale model of a ceiling frame. Identify problems that may be encountered when constructing the frame.
3. Define "stub joists." Explain how to lay out the ends of stub ceiling joists to match the roof pitch.
4. Discuss the methods for fastening partitions that run parallel to the joists to the ceiling frame. Stress that the primary purpose of fastening the partitions to the ceiling frame is to provide support.
5. Have students refer to the local building code to determine the minimum size of the attic scuttle hole. Ask the students to provide ideas on how to frame the scuttle hole.

STRONGBACKS
1. Discuss the purpose of a strongback. Demonstrate how strongbacks are constructed.

ESTIMATING MATERIALS
1. Using a stock plan, determine the wall and ceiling framing members that are required. Suggest that the students mark each wall and partition that they are adding to make sure that all of the walls and partitions are included.
2. Using the same stock plan, determine the number of studs required if they are spaced 16″ O.C.
3. Estimate the amount of wall sheathing required for the house shown in the stock plan.

UNIT REVIEW
1. Review the unit objectives. Be sure that the students fully understand each objective.
2. Assign *Important Terms*, *Test Your Knowledge* questions and *Outside Assignments* on page 231 of the text. Review the answers in class.
3. Assign pages 55-62 of the workbook. Review the answers in class.

EVALUATION
1. Use Procedure Checklist—*Constructing Wall Sections,* to evaluate the techniques that students used to perform these tasks.
2. Use Procedure Checklist—*Erecting Wall Sections,* to evaluate the techniques that students used to perform these tasks.
3. Use Unit 9 Quiz for in-class evaluation. Correct the quizzes and return them to the students for review.

ANSWERS TO TEST YOUR KNOWLEDGE TEXT PAGE 231
1. Bottom member of wall frame which rests on floor.

2. Seven.
3. header
4. Full size guide, marking height of every horizontal member of a wall frame.
5. regular stud
6. story pole
7. sole plate
8. side
9. Wood bracing must be let into the studs by notching. Metal bracing usually requires no more than a saw kerf.
10. 4′
11. cannot
12. rafters
13. walls, partitions
14. E. All of the above.
15. Exterior walls are wrapped to seal cracks and waterproof walls.
16. Welding and fastening with self-tapping screws.

ANSWERS TO WORKBOOK QUESTIONS
PAGES 55-62

1. C. headers
2. B. 16″
3. A. Top plate.
   B. Header.
   C. Trimmer.
   D. Rough sill.
   E. Sole plate.
   F. Cripple stud.
   G. Full stud.
4. B. 10d
5. See Figure 9-6, views A and B, in text.
6. C. 4′-9″
7. C. 1/2″
8. C. 2 × 10
9. D. Lay out two stud positions at each side of all openings.
10. ceiling heights
11. C. distance from rough floor to ceiling
12. C. bottom of header to top of rough sill
13. A. 16d.
    B. 16d.
    C. 10d.
    D. 8d.

14. plumb line
15. A. P.E.T.
16. wall backing
17. B. 1 × 4.
18. B. 16″
19. See Figure 9-38 in text.
20. A. 25/32″
    B. 1 1/2″
21. A. 2″
    B. 8d.
22. C. R-5
23. ceiling joists
24. See Figure 9-53 in text.
25. When it is not load-bearing or when roof trusses are used to carry the roof load.
26. A. 3/4″ weld.
    B. Joist.
    C. 3/4″ weld.
    D. Stud.
    E. 3/4″ weld.
27. Steel framing.
    Load-bearing header.
28. 891′
    bd. ft.: 594
29. studs: 305
30. pcs.: 39

ANSWERS TO UNIT 9 QUIZ

1. True.
2. False.
3. B. 2 × 4
4. D. story pole
5. A. Plywood.
6. C. strongback
7. B. Trimmer
8. F
9. C
10. A
11. D
12. E
13. B

Reproducible Master 9-1
Framing Corners

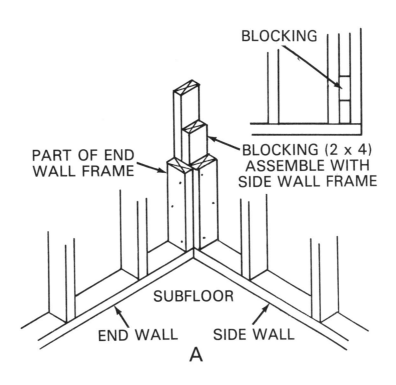

A

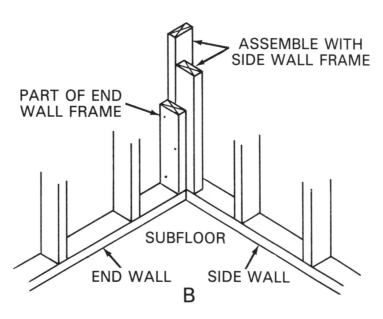

B

Copyright Goodheart-Willcox Co., Inc.

Reproducible Master 9-2
Framing Wall Intersections

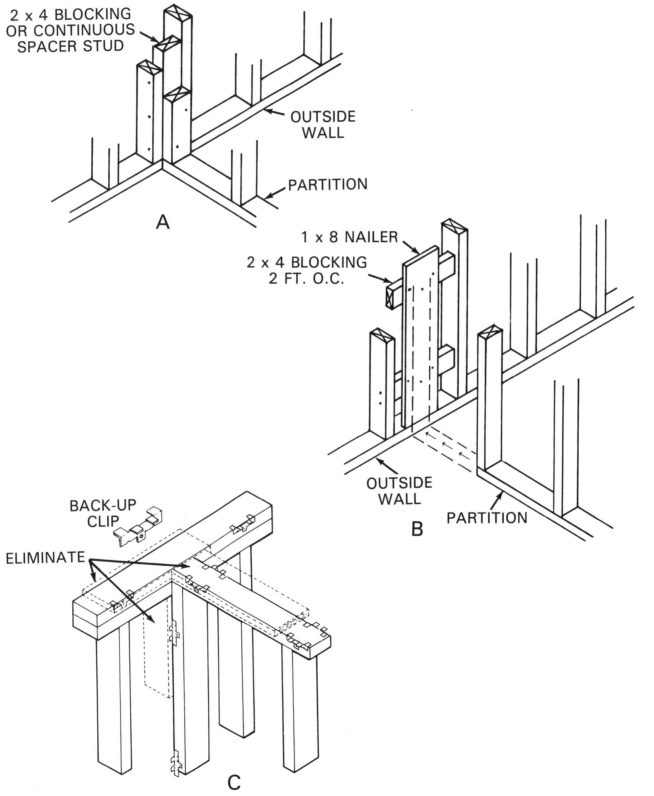

2 x 4 BLOCKING
OR CONTINUOUS
SPACER STUD

OUTSIDE
WALL

PARTITION

A

1 x 8 NAILER

2 x 4 BLOCKING
2 FT. O.C.

OUTSIDE
WALL

PARTITION

B

BACK-UP
CLIP

ELIMINATE

C

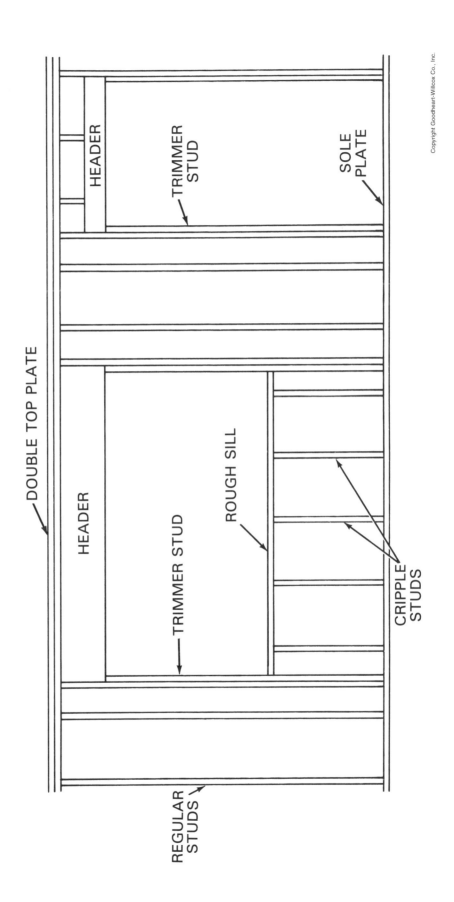

Reproducible Master 9-3

Framing Wall Openings

Reproducible Master 9-4 (Part A)

Framing with Patterns

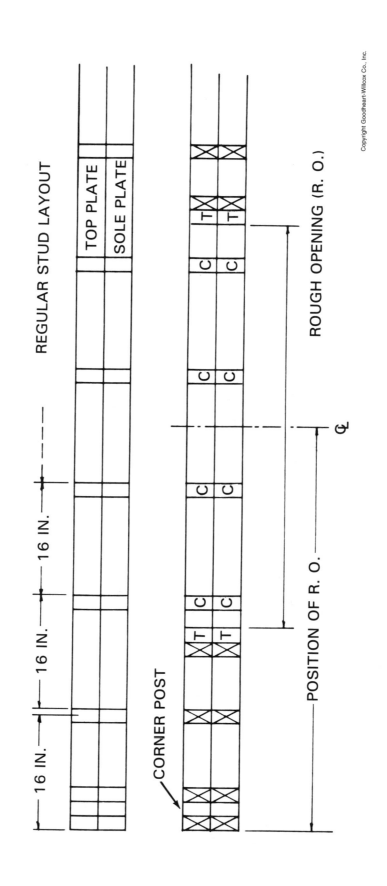

REGULAR STUD LAYOUT

TOP PLATE

SOLE PLATE

16 IN.

16 IN.

16 IN.

16 IN.

CORNER POST

T C

T C

C

C

C

C

C T

C T

ROUGH OPENING (R. O.)

POSITION OF R. O.

C̵L

Reproducible Master 9-4 (Part B)

Framing with Patterns

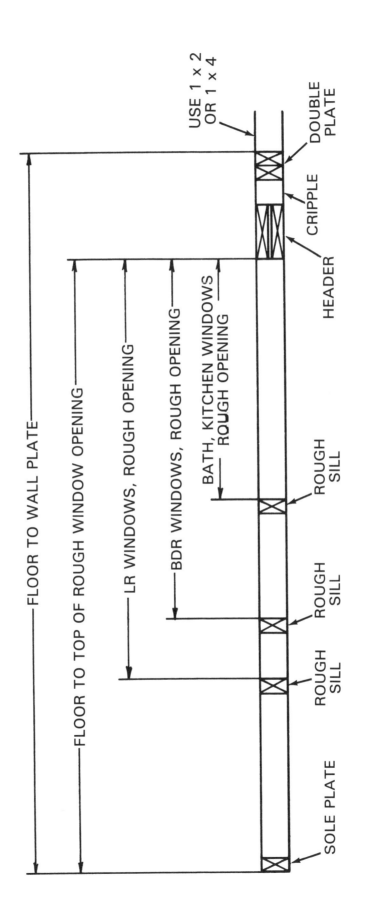

STORY POLE

Framing with Patterns

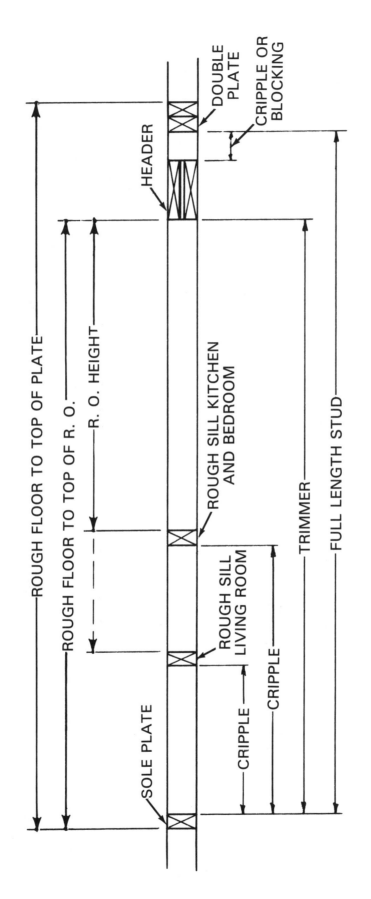

MASTER STUD
PATTERN

Constructing Wall Sections

Name _____ Total _____

	5	4	3	2	1	
❏ Easily selects the appropriate materials for the wall section after referring to the prints.	5	4	3	2	1	Must be told the appropriate materials to use for wall sections.
❏ Readily determines the size of the headers, and cuts and assembles them efficiently.	5	4	3	2	1	Cannot determine the size of the headers; needs help assembling them.
❏ Properly constructs wall section; accurately places studs, making sure crowns are facing up.	5	4	3	2	1	Miscalculates position of studs; disregards position of crowns.
❏ Positions trimmer studs on correct side of full length studs; installed header solidly.	5	4	3	2	1	Places trimmer studs on wrong side of full length studs; header loosely attached.
❏ Accurately lays out position of window openings; determines best time to install cripple studs.	5	4	3	2	1	Cannot determine the position of window openings; not able to install the rough sills and cripple studs.
❏ Determines placement of studs or blocking at partition intersections; installs wall bracing when necessary.	5	4	3	2	1	Not able to determine locations of partition intersections; forgets to check for bracing installation.
❏ Checks squareness of wall section; applies sheathing efficiently, and in the correct direction.	5	4	3	2	1	Neglects to check wall squareness; applies sheathing in the wrong direction.

Unit 9 Procedure Checklist
Erecting Wall Sections

Name _____ Total _____

❏ Determines the best means for raising wall section; readies bracing; makes sure enough workers are available; prepares platform so section doesn't fall off it.	5	4	3	2	1	Neglects to properly prepare work area for raising wall section; tries to erect wall section by self, or with too few workers.
❏ Erects wall section; immediately secures it with braces; adjusts sole plate, and nails it into floor frame.	5	4	3	2	1	Carelessly erects wall section; neglects to brace wall section when erected; does not properly attach sole plate to floor frame.
❏ Carefully loosens braces one at a time and plumbs wall.	5	4	3	2	1	Loosens bracing all at once; does not plumb wall section.

Unit 9 Quiz

Wall and Ceiling Framing

Name _____ Score _____

True-False

Circle T if the answer is True or F if the answer is False.

T   F   1. When listing the size of a rough opening, the width dimension is given first.

T   F   2. Nonbearing partitions require headers.

Multiple Choice

Choose the answer that correctly completes the statement. Write the corresponding letter in the space provided.

_____   3. Studs are generally _____ stock.

   A. 1×4
   B. 2×4
   C. 1×6
   D. 2×6

_____   4. A _____ is a measuring stick that represents the actual wall frame, with markings made at the proper height for every horizontal member of the wall frame.

   A. trimmer
   B. master stud
   C. double header
   D. story pole

_____   5. Which of the following are used as wall sheathing?

   A. Plywood.
   B. Fiberboard.
   C. Composite board.
   D. All of the above.

_____   6. A _____ is an L-shaped support that is attached to the top of joists to strengthen them.

   A. joist hanger
   B. girder
   C. strongback
   D. header

_____   7. _____ studs stiffen the sides of an opening and bear the direct weight of the header.

   A. Cripple
   B. Trimmer
   C. Header
   D. Master

Identification

Identify the parts of the wall frame.

_____ 8. Full stud.

_____ 9. Rough sill.

_____ 10. Header.

_____ 11. Sole plate.

_____ 12. Cripple stud.

_____ 13. Trimmer.

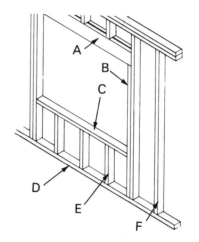

10

Roof Framing

OBJECTIVES

Students will be able to:

❑ List and describe the various types of roofs.
❑ Identify the parts of a common rafter.
❑ Define slope and pitch.
❑ Use a framing square, quick square, and rafter tables.
❑ Lay out common rafters.
❑ Describe the layout and erection of a gable roof.
❑ Explain the design and erection of trusses.
❑ Describe the procedure for sheathing a roof.
❑ Estimate roofing materials.

INSTRUCTIONAL MATERIALS

Text: Pages 233-271
    Important Terms, page 269
    Test Your Knowledge, page 269
    Outside Assignments, page 270
Workbook: Pages 63-72
Instructor's Manual:
    Reproducible Master 10-1, *Roof Types*
    Reproducible Master 10-2, *Plan View of Rafters*
    Reproducible Master 10-3, *Rafter Layout Parts
        & Terms, Parts A* and *B*
    Reproducible Master 10-4, *Rafter Layout, Parts
        A and B*
    Reproducible Master 10-5, *Truss Rafter
        Construction, Parts A* and *B*
    Procedure Checklist—*Erecting a Gable Roof*
    Procedure Checklist—*Erecting Jack Rafters* and
        Installing Fascia
    Unit 10 Quiz
    Section 2 Exam

TRADE-RELATED MATH

The Pythagorean theorem principle is used to calculate the rafter length. The Pythagorean theorem states that if the length any two sides of a right triangle are known, the other side can be determined mathematically. The formula used is:

$$A^2 + B^2 = H^2$$

where A is the altitude (height), B is the base length, and H is the hypotenuse.

This formula can be applied to rafter layout. The altitude can be compared to the rise, or the distance the rafter extends upward above the wall plate. The base length is called the run, or the distance from the outside of the plate to a point directly below the center of the ridge. The hypotenuse can be compared to the line length of the rafter. To determine the line length of the rafter using the previous formula when the rise is 5′ and run is 12′, the following computation is performed:

$$A^2 + B^2 = H^2$$
$$(5)^2 + (12)^2 = H^2$$
$$25 + 144 = H^2$$
$$169 = H^2$$
$$\sqrt{169} = H$$

13 feet is the line length of the rafter

INSTRUCTIONAL CONCEPTS AND STUDENT LEARNING EXPERIENCES

ROOF TYPES

1. Using Reproducible Master 10-1, *Roof Types*, identify the different roof types found in residential construction. Have students identify the types of roofs on their houses and/or apartment buildings.
2. Discuss the advantages (or disadvantages) of each type of house.
3. Have students research the different types of roofs and explain their histories.

ROOF SUPPORTS

1. Review the construction that was discussed in the

last unit. Have students identify the parts of the structure that support the roof.

PARTS OF ROOF FRAME

1. Using Reproducible Master 10-2, *Plan View of Rafters,* discuss the various components of a roof frame. Once again, stress the importance of using the proper terminology to communicate efficiently with other tradesworkers.
2. Using part *A* of Reproducible Master 10-3, *Rafter Layout Parts & Terms,* identify the parts of a rafter and its associated terms. Describe the purpose of each part.

LAYOUT TERMS AND PRINCIPLES

1. Cut a piece of plywood into the shape of a right triangle. Label the hypotenuse, altitude, and base. Describe how you can solve for the third value using the Pythagorean theorem if the other two values are known.
2. Compare the right triangle to the rafter parts shown in part *B* of Reproducible Master 10-3, *Rafter Layout Parts & Terms.*
3. Discuss the difference between the terms "slope" and "pitch." Given the rise and run of several roofs, have the students calculate the slope and pitch of those roofs.

UNIT MEASUREMENTS

1. Review the parts of the framing square, and its use.

FRAMING PLANS

1. Discuss how carpenters lay out roof framing on simple structures.
2. Using a framing plan from a house, have the students determine the type of roof framing that is required.
3. Using the floor plan of a house, have students develop a roof framing plan. Be sure that the students include ridges, overhangs, and all rafters. Stress that the sketch should be as accurate as possible.

RAFTER SIZES

1. Have students refer to the local building code to determine whether rafter sizes are included. If they are included, use the information to determine the rafter sizes for a small residential structure. If they are not included, use Figure 10-8 and a stock plan. Determine the necessary rafter sizes for the structure.

LAYING OUT COMMON RAFTERS

1. Have the students identify the two methods used to lay out common rafters. Briefly discuss both methods.
2. Using parts *A* and *B* of Reproducible Master 10-4, *Rafter Layout,* discuss the use of the framing square or quick square. When students understand the concepts you have presented, demonstrate their application using an actual framing square or quick square. Stress the importance of accurately measuring the dimensions.

USING THE RAFTER TABLE

1. Demonstrate the use of rafter tables on a framing square. Identify the various numbers in the rafter table and explain their meaning.
2. Have the students divide into groups, and give each group a framing square during this activity. Read off the unit rise and rafter length, and have the students determine the run.
3. Using the results from one of the previous activities, have the students use these numbers to layout a pattern rafter. Stress that measurements must be accurate since the pattern rafter will be used to lay out other rafters.

ERECTING A GABLE ROOF

1. Discuss the procedure for assembling a roof frame. Emphasize the importance of teamwork when assembling the frame.
2. Obtain a few framing anchors and clips and pass them out to the students. Explain how the anchors or clips are attached to rafters and plates.
3. Demonstrate the proper procedure for erecting a gable roof. Emphasize that the rafters should be installed with the crown turned upward. Identify places where the rafters should be nailed.

GABLE END FRAME

1. Describe the procedure for constructing the gable end frame of a structure. Demonstrate this procedure by constructing a scale model at ground level. This allows all students to get a close look at the construction techniques.
2. Discuss the various techniques used to construct a gable overhang (shown in Figure 10-24). With the scale model completed in the previous step, construct a gable overhang using one of the methods commonly used in your area.
3. Explain the risk factor involved when working at heights. Stress the importance of using solid

scaffolding and avoiding working directly over someone.

HIP AND VALLEY RAFTERS

1. Identify the types of rafters that are generally included in hip roofs or intersecting gable roofs. Make sure that the students understand the terms: common, hip, valley, jack, and valley rafters before proceeding.
2. Discuss a common procedure for constructing a hip roof. If possible, demonstrate this procedure.
3. Have the students determine the unit run of the hip rafters for a given plan. Explain that 17″ can be used as the unit run for the hip rafters when the unit run of the common rafters is 12″.
4. Have the students explain what is meant by the term "hip jack rafter," and where they are used. Emphasize that if hip jack rafters are evenly spaced that the change in length is consistent from one rafter to another (otherwise known as the common difference).
5. Using a framing square or a quick square, demonstrate how the common differences can be determined from rafter tables. Give different slopes of roofs spaced at different standard O.C. intervals and have students determine the common differences.
6. Discuss how jack rafters are laid out and demonstrate this procedure.
7. Have the students explain the procedure for laying out a valley jack rafter. Make sure that their layout begins at the building line.
8. Demonstrate how jack rafters are cut. Review the safe use of the radial arm saw, and the advantages of using it for cuts of this nature. Stress the importance of tight fitting joints when constructing a roof frame.

ERECTING JACK RAFTERS

1. Discuss the procedure for erecting jack rafters. Explain different techniques that can be used to ensure that rafters are not bowed when applying the sheathing.
2. Have the students explain the purpose of a fascia. Discuss different techniques used to attach the fascia to the vertical ends of the rafters.

SPECIAL PROBLEMS

1. Describe special problems that may occur when framing intersecting roofs. Have the students determine the length of valley cripple jacks.

ROOF OPENINGS

1. Have student identify applications for roof openings. Discuss the procedures for framing both large and small openings.

ROOF ANCHORAGE

1. Identify various means of anchoring a roof. Discuss advantages (or disadvantages) of each anchoring device.

COLLAR BEAMS

1. Discuss the purpose of collar beams. Stress that they do not provide any support, but rather they provide bracing and stiffening of the roof frame. Mention that not every pair of rafters requires a collar beam.

PURLINS

1. Describe the purpose of purlins in residential construction. Explain how purlins are attached to the roof frame.

DORMERS

1. Define "dormer," and discuss the purpose(s) of it.
2. Describe the different types of dormers, including the shed dormer and gable dormer. Explain how each is constructed.

FLAT ROOFS

1. Compare the construction of a flat roof frame to a floor frame. Discuss possible problems that may occur with a roof that is perfectly flat in cold weather locations.
2. Describe the construction of a flat roof. Have students take note of the increased size of the roof joists due to the increase in combined load.

GAMBREL ROOF

1. Discuss the advantages of a gambrel roof in residential construction. Identify the angles commonly used for the upper roof surface (20°) and the lower roof surface (70°).
2. Have the students make a full-size sectional drawing of the intersection between the two slopes. Stress the advantage of taking the time to draw this sectional drawing, rather than trying to construct the roof without it.

SPECIAL FRAMING

1. Using Figures 10-58 and 10-59, discuss some of the special framing problems that may be encountered

ROOF TRUSS CONSTRUCTION

1. Have the students describe the purpose of a roof truss. Discuss the advantages of using roof trusses for standard roof frames.
2. Using part *A* of Reproducible Master 10-5, *Truss Rafter Construction,* identify the common parts of a truss rafter. State that the carpenter is seldom responsible for determining the sizes of truss members, but he/she should understand the principles involved in their construction.
3. Obtain plates and connectors that are commonly used in truss construction. Using part *B* of Reproducible Master 10-5, *Truss Rafter Construction,* identify each of the plates or connectors.
4. Discuss the most efficient method of constructing truss rafters. Demonstrate their construction by creating a scale model of a pattern. When the pattern has been completed and checked for accuracy, have the students construct other scale model trusses.
5. Have students obtain information about portable truss assembly units, and discuss the advantages of using such equipment.
6. Describe the means used to install roof trusses. Stress that carpenters are usually responsible for installing the trusses, and that much of the work involves working at heights.
7. Discuss the types of bracing that are commonly used to support roof trusses.

ROOF SHEATHING

1. Discuss the purpose of roof sheathing. Identify the various materials used for roof sheathing.
2. List the steps necessary to prepare for applying roof sheathing, including setting up scaffolding and moving the sheathing materials into a position where they can be easily reached.
3. Discuss the application of board sheathing when the finish roofing material is to be wood shingles, metal sheets, or tile. Stress that the joints must be made over the center of rafters.
4. Have the students identify the advantages of using structural panels for roof sheathing. Identify the thicknesses commonly used for the different types of finish roofing material.
5. Discuss problems that may occur when using panel materials. As a class, design a ladder rack that can be used to hold roof sheathing.

PANEL CLIPS

1. Obtain panel clips to show to the students. Have them describe where these clips should be used.

ESTIMATING MATERIALS

1. Discuss the method used to determine the number of rafters required for a plain gable roof.
2. Discuss the method used to estimate the number of rafters required for a hip roof.
3. Discuss the means for estimating the amount of sheathing needed for a roof. Mention that a waste factor should always be included to make sure that there is enough material on the job site.

MODEL CONSTRUCTION

1. Throughout the chapter, it has been suggested that scale models be used to demonstrate the construction concepts used for roof framing of a typical residential structure. Scale models can be constructed to save money, yet still convey the construction methods used on the job. The most important concepts that students should gain from the scale models are the techniques used to construct them.

UNIT REVIEW

1. Review the unit objectives. Be sure that the students fully understand each objective.
2. Assign *Important Terms, Test Your Knowledge* questions and *Outside Assignments* on pages 269-271 of the text. Review the answers in class.
3. Assign pages 63-72 of the workbook. Review the answers in class.

EVALUATION

1. Use Procedure Checklist—*Erecting a Gable Wall,* to evaluate the techniques that students used to perform these tasks.
2. Use Procedure Checklist—*Erecting Jack Rafters,* to evaluate the techniques that students used to perform these tasks.
2. Use Unit 10 Quiz for in-class evaluation. Correct the quizzes and return them to the students for review.
3. Use the Section 2 Exam to evaluate the students' knowledge of information found in Units 7-10 of the text.

ANSWERS TO TEST YOUR KNOWLEDGE TEXT PAGE 269

1. hip
2. span
3. 16
4. odd unit
5. seat
6. ridge

7. intersection
8. rake
9. 16′ and 9 7/8″, 21′ and 10 5/8″
10. sixth
11. common
12. pairs
13. collar beams
14. shed
15. purlin
16. camber
17. strength/rigidity
18. end-matched
19. rafter spacing
20. common rafter
21. See page 263.
22. With self-tapping screws.

ANSWERS TO WORKBOOK QUESTIONS
PAGES 63-72

1. A. Shed.
   B. Hip.
   C. Gable.
   D. Flat.
   E. Gambrel
2. Mansard
3. common
   hip
   hip jack
4. A. Plumb cut.
   B. Bird's-mouth.
   C. Tail/overhang.
   D. Tail cut.
5. A. Line length of rafter.
   B. Overhang.
   C. Run.
   D. Span.
   E. Rise.
6. A. rise
   B. run
7. 24
8. face
9. D. 1/12
10. C. laying out full units
11. 8′-11 1/2″
12. pitch
13. rafter tables
14. 7″
15. See Figure 10-18 in text.
16. A. Roof slope.
17. C. 17″
18. 14′-9 1/4″

19. B. ledger
20. A. Hip rafter.
    B. Valley rafter.
21. C. one-half of 45° thickness of ridge
22. A. Common difference.
    B. Unit rise.
    C. Rafter spacing.
23. A. common rafter pattern
24. C. fifth
25. equal
26. 2″
27. dormer
28. collar beams
29. roof joists
30. A. Top chord.
    B. Bottom chord.
    C. Tension web.
    D. Compression web.
31. C. 1/2″
32. A. Kingpost
    B. Scissors.
33. A. Ceiling joist.
    B. Cantilevered support.
    C. Ribbon.
    D. Lookout.
34. nailing base
35. A. 1/2″
36. C. 1/2″
37. 74
38. A. 14′
    B. 44
    C. 616 bd. ft.
39. A. 48 pcs.
    B. 20′
40. A. 56 pcs.
    B. 1792 sq. ft.
41. A. 78 pcs.
    B. 2485 sq. ft.
42. 17 squares

ANSWERS TO UNIT 10 QUIZ

1. False.
2. True.
3. True.
4. F
5. C
6. E
7. D
8. G
9. A
10. B

11. H
12. C
13. D
14. A
15. B
16. B. Hip
17. C. 1/8
18. A. Hip jack

ANSWERS TO SECTION 2 EXAM

1. False.
2. False.
3. False.
4. False.
5. True.
6. True.
7. False.
8. True.
9. True.
10. True.
11. B. hydration
12. C. areaway
13. D. 3/8″
14. C. 9.5
15. A. 4″
16. D. All of the above are admixtures.
17. D. All of the above.
18. B. Trimmer studs
19. B. ceiling joists
20. B
21. C
22. A

23. B
24. D
25. C
26. E
27. A
28. A
29. E
30. B
31. D
32. C
33. ledger
34. 8
35. sixth
36. sump pit
37. 4.5
38. ribbon
39. 1/2
40. P.E.T.
41. E
42. A
43. C
44. D
45. G
46. B
47. F
48. A
49. D
50. B
51. C
52. F
53. E

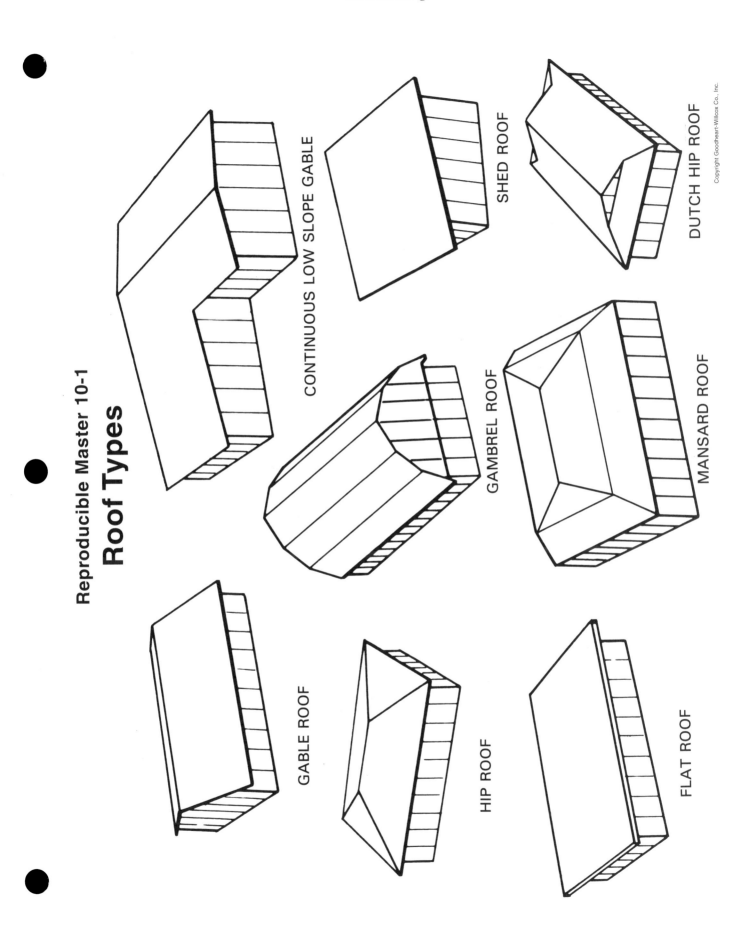

Reproducible Master 10-1

Roof Types

CONTINUOUS LOW SLOPE GABLE

SHED ROOF

DUTCH HIP ROOF

GAMBREL ROOF

MANSARD ROOF

GABLE ROOF

HIP ROOF

FLAT ROOF

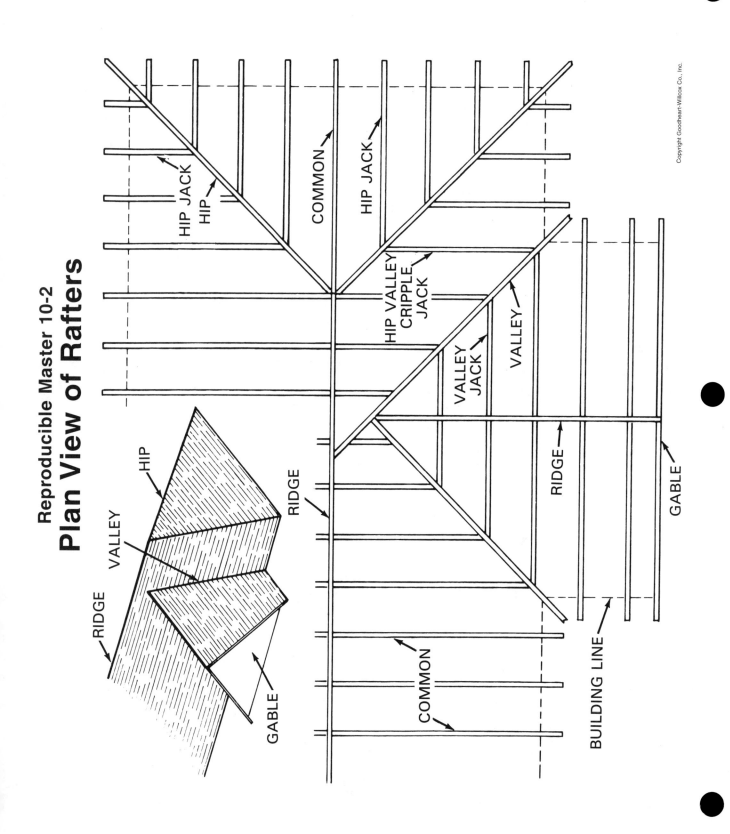

Reproducible Master 10-2
Plan View of Rafters

HIP JACK
HIP
COMMON
HIP JACK
HIP VALLEY CRIPPLE JACK
VALLEY JACK
VALLEY
RIDGE
GABLE
BUILDING LINE
COMMON
RIDGE
RIDGE
HIP
VALLEY
RIDGE
GABLE
COMMON

Reproducible Master 10-3 (Part A)
Rafter Layout Parts & Terms

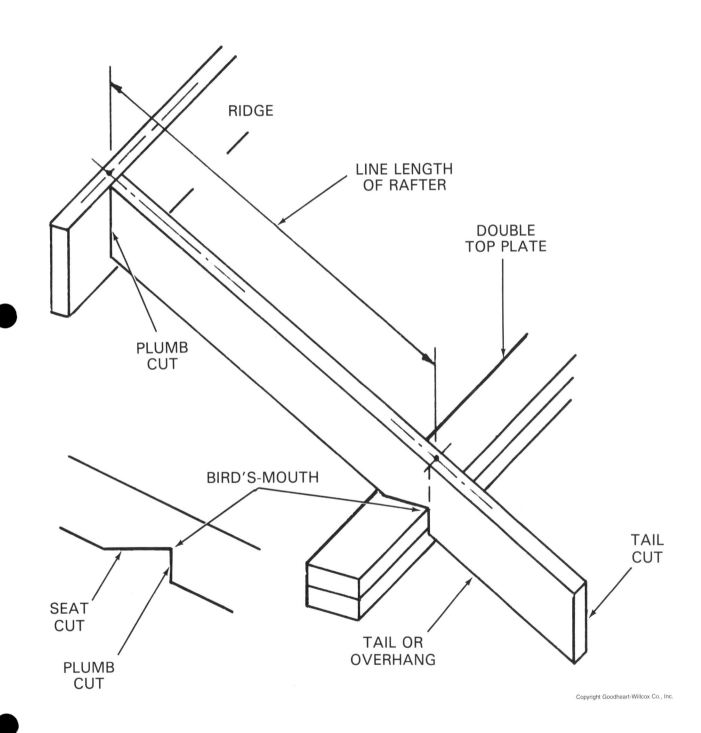

RIDGE

LINE LENGTH
OF RAFTER

DOUBLE
TOP PLATE

PLUMB
CUT

BIRD'S-MOUTH

TAIL
CUT

SEAT
CUT

PLUMB
CUT

TAIL OR
OVERHANG

Reproducible Master 10-3 (Part B)

Rafter Layout Parts & Terms

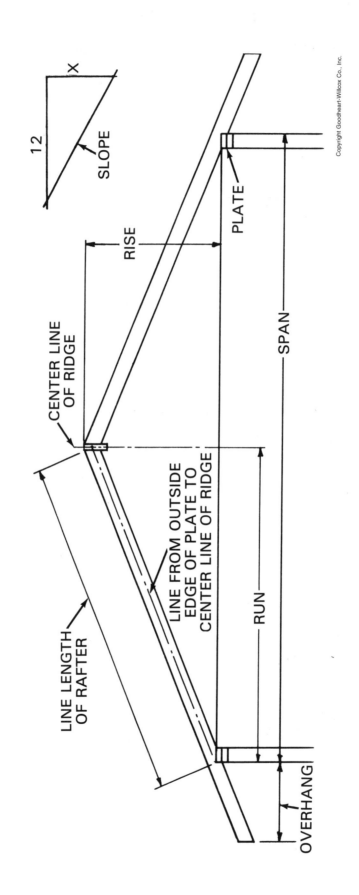

Reproducible Master 10-4 (Part A)
Rafter Layout

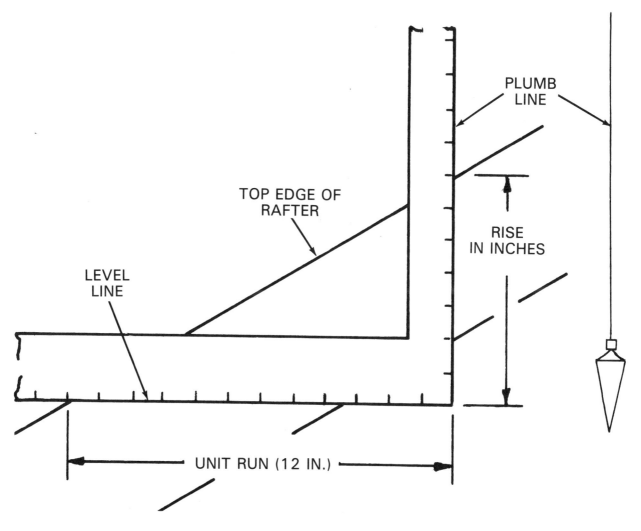

PLUMB LINE

TOP EDGE OF RAFTER

RISE IN INCHES

LEVEL LINE

UNIT RUN (12 IN.)

Rafter Layout

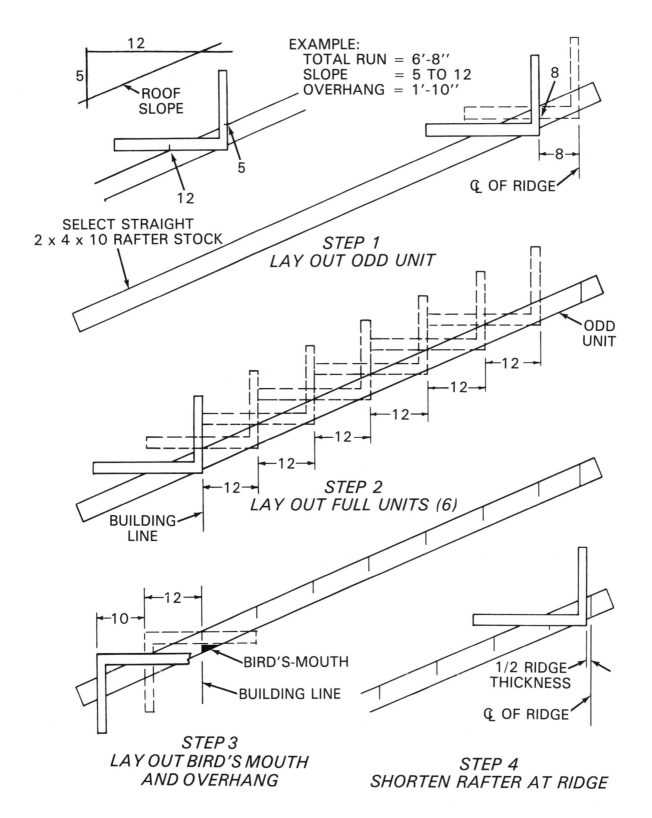

12

5

ROOF
SLOPE

5

12

SELECT STRAIGHT
2 x 4 x 10 RAFTER STOCK

EXAMPLE:
 TOTAL RUN = 6'-8''
 SLOPE   = 5 TO 12
 OVERHANG = 1'-10''

8

8

℄ OF RIDGE

STEP 1
LAY OUT ODD UNIT

ODD
UNIT

12

12

12

12

12

12

BUILDING
LINE

STEP 2
LAY OUT FULL UNITS (6)

12

10

BIRD'S-MOUTH

BUILDING LINE

STEP 3
LAY OUT BIRD'S MOUTH
AND OVERHANG

1/2 RIDGE
THICKNESS

℄ OF RIDGE

STEP 4
SHORTEN RAFTER AT RIDGE

Reproducible Master 10-5 (Part A)

Truss Rafter Construction

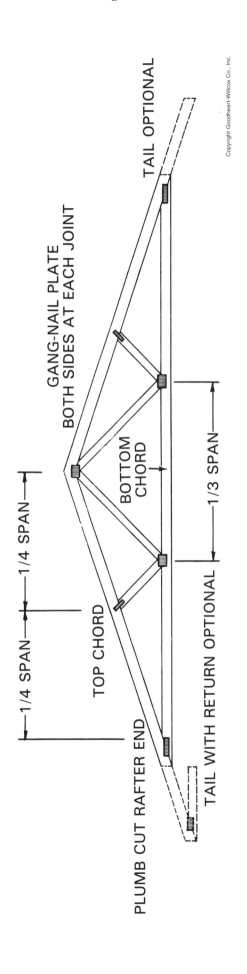

GANG-NAIL PLATE
BOTH SIDES AT EACH JOINT

TAIL OPTIONAL

BOTTOM CHORD

1/3 SPAN

TOP CHORD

1/4 SPAN

1/4 SPAN

PLUMB CUT RAFTER END

TAIL WITH RETURN OPTIONAL

Reproducible Master 10-5 (Part B)

Truss Rafter Construction

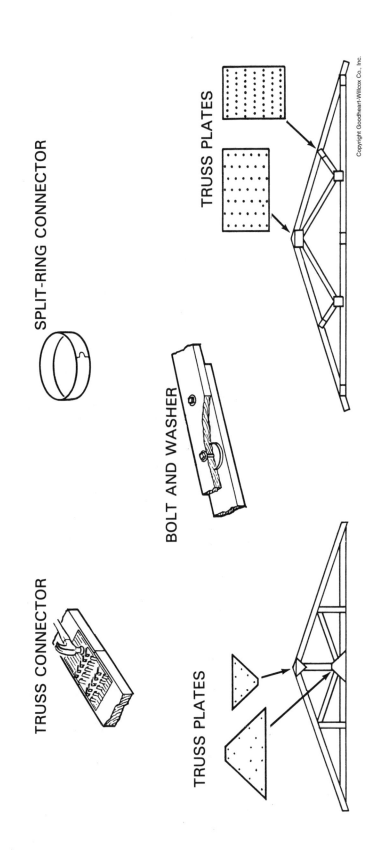

SPLIT-RING CONNECTOR

TRUSS PLATES

BOLT AND WASHER

TRUSS CONNECTOR

TRUSS PLATES

Copyright Goodheart-Willcox Co., Inc.

Unit 10 Procedure Checklist
Erecting a Gable Roof

Name _____ Total _____

❏ Selects appropriate materials for ridge; accurately lays out rafter spacing by transferring markings from the plate or layout rod; makes sure that joints in the ridge will occur at the center of a rafter; carefully cuts ridge members and lays them across ceiling joists.	5	4	3	2	1	Disregards stock's condition when selecting material for the ridge; measures rafter spacing rather than transferring it; neglects to have ridge joints occur at centers of rafters; tries to cut ridge members after moving stock to the ceiling joists.
❏ Makes sure that rafters, ridge boards, and bracing are at hand; recruits other worker(s) to help erect gable; temporarily nails rafter at plate first, then at ridge; moves about five rafters down and installs next rafter, making sure the ridge is level.	5	4	3	2	1	Must constantly be rounding up materials for the roof frame while it is being erected; tries to erect frame by her or himself; installs rafter adjacent to the end rafter; neglects to level ridge.
❏ Installs rafters between main supporting rafters using 16d nails; places only a few rafters on one side before installing matching rafters; installs additional sections of the ridge and assembles the rafters sections; adds bracing where required.	5	4	3	2	1	Uses inappropriate nails when installing rafters; erects rafters all on one side before erecting matching rafters; neglects to use bracing.

Unit 10 Procedure Checklist
Erecting Jack Rafters and Installing Fascia

Name _____   Total _____

❏ Accurately cuts jack rafters; erects jack rafters in pairs to prevent jack and valley rafters from being pushed out of line; nails stock using an appropriate size nail; uses bracing where necessary.	5	4	3	2	1	Miscalculates rafter dimensions, resulting in poor joints or new stock being cut; erects all jack rafters on one side, pushing the jack and valley rafters out of line; uses too small or too large nails; neglects to use bracing if necessary.
❏ Checks over entire roof frame; corrects bows; sights rafters and shims where necessary.	5	4	3	2	1	Neglects to check over roof frame and correct problems.
❏ Determines best method of attaching fascia; accurately cuts the fascia, matching the angle of the upper edge of the fascia to the roof angle; cuts miters on corners; carefully installs fascia using correct size nails.	5	4	3	2	1	Miscalculates length of fascia; neglects to cuts upper edge of fascia to match roof angle; uses butt joints for corners; uses too small or too large nails to attach the fascia.

Unit 10 Quiz
Roof Framing

Name _____ Score _____

True-False

Circle T if the answer is True or F if the answer is False.

T   F   1.   A gable roof has four sloping sides.

T   F   2.   In a plan view, common rafters run at a right angle from the wall plate to the ridge.

T   F   3.   Collar beams tie rafters and the ridge together, reinforcing the roof frame.

Identification

Identify the parts of the rafter illustrated to the right.

_____   4.   Plumb cut.

_____   5.   Double top plate.

_____   6.   Tail or overhang.

_____   7.   Tail cut.

_____   8.   Seat cut.

_____   9.   Ridge.

_____   10.   Line length of rafter.

_____   11.   Bird's mouth.

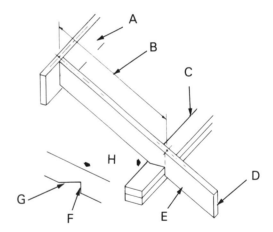

Identify the parts of the rafter illustrated to the right.

_____   12.   Bottom chord.

_____   13.   Return.

_____   14.   Top chord.

_____   15.   Tail.

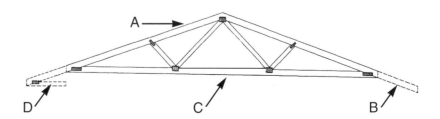

Multiple Choice

Choose the answer that correctly completes the statement. Write the corresponding letter in the space provided.

_____ 16. _____ rafters run from the wall plate to the ridge at a 45° angle.

   A. Common
   B. Hip
   C. Valley
   D. None of the above.

_____ 17. If the total roof rise is 4′ and the total span is 32′, the pitch is _____.

   A. 1/2
   B. 1/6
   C. 1/8
   D. 1/10

_____ 18. _____ rafters run between the wall plate and hip rafter.

   A. Hip jack
   B. Valley
   C. Common
   D. None of the above.

Section 2 Exam
Footing, Foundations, and Framing

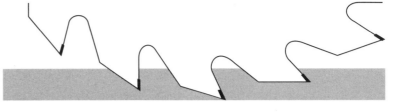

Name _____ Score _____

True-False

Circle T if the answer is True or F if the answer is False.

T   F   1. Platform framing is also referred to as balloon framing.

T   F   2. If large holes are to be cut in a joist for a plumbing run, they should be cut along the bottom edge.

T   F   3. In cold climates, footings should be located above the frost line so that freezing can be avoided.

T   F   4. A key is formed in a foundation wall by cutting it in with a concrete saw.

T   F   5. A first course of concrete block should first be laid up dry to avoid unnecessary cutting.

T   F   6. The compressive strength of concrete is high, while its tensile strength is relatively low.

T   F   7. Perimeter insulation is not required for slab-on-grade construction in cold climates.

T   F   8. Screeding is the process of roughly spreading concrete between forms when it is first placed.

T   F   9. When positioning joists, the crowns should always be turned upward.

T   F   10. Interior, non-load bearing walls are called partitions.

Multiple Choice

Choose the answer that correctly completes the statement. Write the corresponding letter in the space provided.

_____ 11. The chemical action that occurs when cement, aggregate, and water are added together is called _____.
  A. setting
  B. hydration
  C. thermosiphoning
  D. None of the above.

_____ 12. The steel or masonry structure that provides space around basement windows located below finished grade is a(n) _____.
  A. window well
  B. window buck
  C. areaway
  D. lintel

_____ 13. A _____ mortar joint is commonly used with standard concrete blocks.
  A. 1/16″
  B. 1/8″
  C. 1/4″
  D. 3/8″

_____ 14. _____ cubic yards of concrete are required for a slab measuring 6″ × 20′ × 25′. (Round your answer to the nearest half cubic yard.)

A. 6
B. 7.5
C. 9.5
D. 11

_____ 15. In most areas of the country, sidewalks are _____ thick.

A. 4″
B. 6″
C. 8″
D. 10″

_____ 16. _____ are *not* a type of admixture.

A. Accelerators
B. Air-entraining agents
C. Corrosion inhibitors
D. All of the above are admixtures.

_____ 17. Subfloors _____.

A. add rigidity to the structure
B. provides a base for finish flooring material
C. provide a layout surface for carpenters when constructing additional framing
D. All of the above.

_____ 18. _____ stiffen the sides of a wall opening and bear the direct weight of the header.

A. Cripple studs
B. Trimmer studs
C. Columns
D. Posts

_____ 19. The ceiling of a one-story structure is supported by a framework consisting of members called _____.

A. trusses
B. ceiling joists
C. rafters
D. None of the above.

Matching

Select the correct answer from the list on the right and place the corresponding letter in the blank on the left.

_____ 20. Plain footings.

_____ 21. Reinforced footings.

_____ 22. Stepped footings.

A. Changes grade levels at intervals to compensate for a sloping site.
B. Commonly used for light loads.
C. Steel rebar is used for added strength against cracking.

Name _____

_____ 23. Joist.

_____ 24. Sill.

_____ 25. Sole plate.

_____ 26. Joist header.

_____ 27. Girder.

A. Provides support for joists when long spans are necessary.
B. Horizontal framing members used to support floor and ceiling loads.
C. Lowest horizontal member on wall and partition framing.
D. Wall studs are usually tied into this framing member.
E. Outside member that ties joists together.

_____ 28. Common rafter.

_____ 29. Hip rafter.

_____ 30. Valley rafter.

_____ 31. Valley jack rafter.

_____ 32. Cripple jack rafter.

A. Runs at 90° (in the plan view) from the wall plate to the ridge.
B. Extends diagonally from the wall plate to the ridge in the hollow formed by the intersection of two roof sections.
C. Intersects neither the plate nor the ridge, and is terminated at each end by hip and valley rafters.
D. Same as the upper end of the common rafter, but intersects a valley rafter rather than the plate.
E. Same as the lower part of a common rafter, but intersects a hip rafter rather than the ridge.

Completion

Place the answer that correctly completes the statement in the space provided.

_____ 33. A batter board assembly consists of stakes and at least one horizontal member called a(n) _____.

_____ 34. Foundations should generally extend at least _____ inches above finished grade.

_____ 35. Intersecting concrete block walls should be tied into each other every _____ course.

_____ 36. A special water collection basin used to ensure good drainage around a foundation is called a(n) _____.

_____ 37. A slab measuring 4″ × 15′ × 20′ will require _____ cubic yards of concrete. Add 5% waste to allow for waste and slab thickness variations. (Round your answer to the nearest half cubic yard.)

_____ 38. In balloon framing, the horizontal member that is attached to the studs and supports the second floor joists is called the _____.

_____ 39. Headers are formed by nailing two members together with a _____ inch spacer between.

_____ 40. In modern construction, studs are usually cut to length at the mill, and are designated with the letters _____.

Identification

Identify the types of concrete blocks.

_____ 41. Stretcher.

_____ 42. Corner.

_____ 43. Bull nose.

_____ 44. Jamb.

_____ 45. Lintel.

_____ 46. Pier.

_____ 47. Partition.

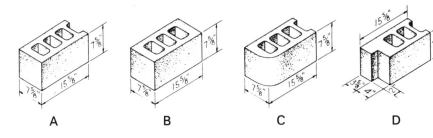

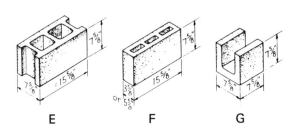

Identify the types of roofs.

_____ 48. Gable roof.

_____ 49. Shed roof.

_____ 50. Hip roof.

_____ 51. Gambrel roof.

_____ 52. Dutch hip roof.

_____ 53. Mansard roof.

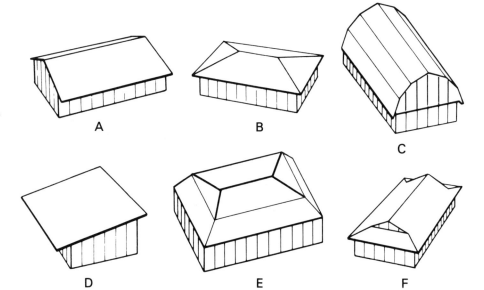

11

Roofing Materials and Methods

OBJECTIVES

Students will be able to:

- ❏ List the covering materials commonly used for sloping roofs.
- ❏ Define roofing terms.
- ❏ Describe how to prepare the roof deck.
- ❏ Describe reroofing procedures for both asphalt and wood shingles.
- ❏ Illustrate correct nailing patterns.
- ❏ Select appropriate roofing materials for various slopes and conditions.
- ❏ Describe the application procedure for a built-up roof.
- ❏ Explain how various roofing products are applied.
- ❏ Demonstrate the proper positioning of gutters.
- ❏ Estimate materials needed for a specific roofing job.

INSTRUCTIONAL MATERIALS

Text: Pages 275-316

  Important Terms, page 314

  Test Your Knowledge, page 315

  Outside Assignments, page 316

Workbook: Pages 73-80

Instructor's Manual:

  Reproducible Master 11-1, *Shingling Valleys, Parts A, B,* and *C*

  Reproducible Master 11-2, *Six Inch Shingling Method*

  Reproducible Master 11-3, *Four Inch Shingling Method*

  Reproducible Master 11-4, *Vent Stack Flashing*

  Reproducible Master 11-5, *Metal Gutter System*

  Procedure Checklist—*Applying Wood Shingles*

  Procedure Checklist—*Installing Asphalt Shingles*

  Unit 11 Quiz

TRADE-RELATED MATH

One method of estimating roofing material is to calculate the total surface area to be covered. In new construction, the figures used to estimate the sheathing can also be used to estimate the underlayment and finished roofing material.

Another method to estimate roof area is to determine the total ground area of the structure including the eaves and cornice overhang. The ground area is converted to the roof area by adding a percentage. These percentages are shown at the bottom of page 313 of the text. In either case, the roof area is then divided by 100 to determine the number of squares of roofing material to be used.

The following formula can be used:

$$\frac{\text{Roof area}}{100} = \text{number of squares}$$

In the following example, the total ground area is 1650 square feet, and the slope of the roof is 5 in 12.

$$\text{Roof area} = 1650 + (1650 \times 8.5\%)$$
$$= 1650 + (1650 \times .085)$$
$$= 1650 + 140.25$$
$$\text{Roof area} = 1790.25$$

$$\frac{\text{Roof area}}{100} = \text{number of squares}$$

$$\frac{1790.25}{100} = 17.9$$

18 squares of roofing material should be purchased

INSTRUCTIONAL CONCEPTS AND STUDENT LEARNING EXPERIENCES

TYPES OF MATERIALS

1. Discuss the factors to consider when selecting roofing materials. Emphasize that a carpenter should always check the local building code before installing any type of roofing material to

make sure that a particular type of material is not prohibited.

2. Have the students identify various types of roofing materials including: shingles (made of a variety of materials), slate, tile, and sheet materials of metal and other materials.

ROOFING TERMS

1. Have the students define the following roofing terms: square, coverage, exposure, head lap, side lap, and shingle butt. Refer to Figure 11-3 to illustrate each term.

PREPARING THE ROOF DECK

1. Discuss the necessary preparation for a roof deck. Identify various things that a carpenter should be aware of when inspecting a roof deck in preparation for the roofing material.

2. Explain the purpose of using louvered openings for an attic.

ASPHALT ROOFING PRODUCTS

1. Have the students list the three general categories of asphalt roofing products, and discuss the purpose of each.

2. Discuss the composition of saturated felts. Mention that saturated felts are commonly used under shingles for sheathing paper.

3. Discuss the composition of ice and water barrier. Be sure to cover its purpose and uses.

4. Discuss the composition of roll roofing and shingles. Explain the purpose of the mineral granules embedded in the surface.

5. Examine the charts on pages 278 and 279 of the text regarding shingles and roll roofing. Have students note the weight and size of each.

6. Using a standard three-tab asphalt strip shingle, measure its dimensions. Dimensions can be written on duct tape placed directly on the shingle.

7. Discuss the advantages of applying self-sealing shingles during the summer.

8. Emphasize the risk factor involved in roofing a structure. Stress the importance of using solid scaffolding when applying shingles.

UNDERLAYMENT

1. Discuss the purpose of underlayment. Identify materials commonly used as underlayment in your specific location. Also, identify materials that should not be used as underlayment.

2. Describe the proper application of underlayment materials.

DRIP EDGE

1. Describe the purpose and proper application of a drip edge.

FLASHING AT EAVES

1. Discuss the purpose of an eaves flashing strip or ice and water barrier in cold climates.

OPEN VALLEY FLASHING

1. Define "flashing," and identify materials that are commonly used as flashing.

2. Using part A of Reproducible Master 11-1, Shingling Valleys, describe the proper application of open valley flashing.

3. Discuss the purpose of snapping a chalk line along the edges and down the center of the valley.

WOVEN AND CLOSED-CUT VALLEYS

1. Discuss the purpose of using woven or closed-cut valleys.

2. Using parts B and C of Reproducible Master 11-1, Shingling Valleys, discuss the preparations necessary before installing a woven or closed-cut valley. Compare and contrast the two types.

FLASHING AT A WALL

1. Discuss the purpose of using metal flashing at a wall and roof intersection. Describe the size of flashing that should be used. Stress that the flashing should not be nailed to the wall since the settling of the roof frame may damage the seal.

STRIP SHINGLES

1. Explain the difference in laying shingles for a small roof and a large roof (over 30′ long).

2. Discuss the need for snapping chalk lines for inexperienced carpenters or roofers. Describe where the chalk lines should be snapped.

3. Since shingles may vary from manufacturer to manufacturer, stress that a carpenter should always read the manufacturer's recommendations before applying shingles.

4. Obtain various types of nails used for installing asphalt shingles and have the students identify each type. Discuss the length of nail that should be used.

5. Describe the number of nails and placement for the correct application of roofing materials. Mention that the shingle should be nailed from one edge to the other to avoid buckling it.

6. If using a pneumatic stapler for shingle application, review the safety precautions.

Whether using nails or staples, stress that they should be driven square and flush to the shingle, and not sunken into the shingle.

7. Explain the purpose of a starter strip and how it should be applied.

8. Using Reproducible Master 11-2, *Six Inch Shingling Method*, describe the procedure for laying shingles. Stress the importance of accurate cuts and exactly aligning the cutouts in every other course to secure a neat appearance.

9. Using Reproducible Master 11-3, *Four Inch Shingling Method*, compare the 4″ and 6″ methods. Point out the differences between the two methods.

10. Demonstrate the six and four inch shingling methods using a 4′ × 8′ sheet of plywood mounted on an angled support set up at table top level. (This will allow all students to see the techniques involved in laying shingles without exposing them to heights.) Be sure to treat the sheet of plywood as if it were an entire roof section. Apply underlayment, drip edges, and a starter strip for realism.

CHIMNEY FLASHING

1. Identify the two parts of chimney flashing and discuss why both are required.

2. Describe the correct procedure for applying flashing around a chimney that does not have a saddle.

CHIMNEY SADDLE

1. Discuss the construction of a chimney saddle. Identify the materials commonly used for the covering of chimney saddles (sheet metal or roll roofing).

VENT STACK FLASHING

1. Using Reproducible Master 11-4, *Vent Stack Flashing*, discuss the procedure for installing flashing around a vent stack.

2. Remove a portion of the shingles on the simulated roof used to show shingling methods. Drill (or cut) a hole in this roof to accept a vent stack. Secure the vent stack into position on the underside of the plywood. Demonstrate the correct procedure for installing vent stack flashing.

HIPS AND RIDGES

1. Describe how hip and ridge shingles can be fabricated from common square-butt shingle

strips or from mineral surfaced roll roofing. Demonstrate this procedure as well. Discuss how the hip or ridge shingles are attached to the roof.

WIND PROTECTION

1. Review the purpose of the adhesive strips found on some shingles. Stress that self-sealing shingles are generally satisfactory only for roofs with slopes up to 60°.

2. Discuss procedures that should be followed for installing roofing materials on steep roofs, such as mansard roofs.

INDIVIDUAL ASPHALT SHINGLES

1. Refer to Figure 11-25 and discuss the use of individual asphalt shingles for roofing. Cite advantages and disadvantages of using this type of shingle.

LOW-SLOPE ROOFS

1. Discuss special procedures that should be followed for low-slope roofs.

ROLL ROOFING

1. Point out the two parts of typical roll roofing—the granular surfaced area and the selvage.

2. Describe the proper procedure for applying roll roofing.

3. Remove the asphalt shingles installed on the simulated roof. Demonstrate how roll roofing is applied to a roof and cemented into place.

REROOFING

1. Discuss the two primary factors involved in determining whether old roofing should be removed before applying new roofing. Discuss the means for removing old shingles if it is determined that they must be removed.

2. Explain the steps taken to reroof over old wood shingles. Emphasize the importance of repairing and renailing existing shingles if necessary.

3. Discuss the preparation necessary for reroofing over old asphalt shingles.

4. Describe steps that should be taken to seal the joint between a vertical wall and roof surface in a reroofing operation.

BUILT-UP ROOFING

1. Describe situations in which built-up roofing is advantageous over asphalt shingles.

2. Discuss the procedure for applying built-up roofing materials. Explain the purpose of

spreading slag, gravel, crushed stone, or marble chips over the surface of the built-up roof.

3. Compare the terms "dead-flat asphalt" and "steep asphalt." Discuss the importance of using the correct type of asphalt for the situation.

4. Describe the purpose of a gravel stop and discuss its installation on a built-up roof.

5. Identify the types of precautions that must be taken around chimneys, vents, and the intersections of the walls and the roof. Describe the basic construction commonly used at the intersection of a flat roof and wall.

WOOD SHINGLES

1. Discuss the advantages and disadvantages of using wood shingles for roofing.

2. Have the students identify the types of wood commonly used for wood shingles. Explain why these types of wood are more predominant than other types.

3. Explain that the exposure of wood shingles depends on the slope of the roof. List different exposures that are used for different sizes of shingles.

4. Identify the types of sheathing that are commonly used for wood shingles. Discuss why one type of sheathing might be used rather than another.

5. Describe the types of underlayment that can be used for wood shingles, including fire-resistant underlayment.

6. Explain that flashing may be required in areas where temperatures drop below 0°F (–19°C). Identify the types of flashing, including ice and water barrier. Describe the proper application of each.

7. Discuss why only rust-resistant nails should be used with wood shingles. Identify the types of nails commonly used for wood shingles by referring to Figure 11-40.

8. Discuss the proper use of a shingler's hatchet when installing wood shingles and shakes. Stress safety precautions since the blade and heel are sharp.

9. Describe the procedure for applying the first course of shingles. Stress the nailing pattern that should be used, and that the shingles should not be crushed when installing the nails. Also mention that the shingles should be spaced about 1/4″ apart to allow for expansion.

10. Explain the installation of the second and successive rows of wood shingles. Describe why a straightedge is used when installing wood shingles.

11. Demonstrate the installation of wood shingles on the simulated roof. Emphasize that alignment of the shingles should be checked every fifth or sixth course to ensure that the courses run parallel with the ridge. Show the proper means of installing a piece of beveled siding along the roof edge to tilt the shingles inward.

12. Discuss the procedure of installing wood shingles at hips and ridges using prefabricated units and common wood shingles.

13. Discuss the types of special effects that can be obtained with wood shingles. However, stress that the basic procedure for installing wood shingles should be accomplished before trying to create special effects.

14. Discuss the preparation necessary for reroofing with wood shingles. Describe how new wood shingles are installed over an existing wood shingled roof.

WOOD SHAKES

1. Discuss the similarities and differences between wood shakes and wood shingles.

2. Point out that the recommended minimum slope for roofs on which shakes are to be applied is 4 in 12. Discuss the exposure necessary for various sizes of wood shakes.

3. Describe the preparation necessary for wood shakes.

4. Discuss the installation of wood shakes. Stress that there should be 1/4″ to 3/8″ of space between shakes and that the joints of successive courses should be offset at least 1 1/2″.

5. Review the types of nails that are used for wood shingles. Mention that these are the same types of nails used for wood shake installation.

6. Compare the installation of wood shingles to that of wood shakes.

7. Remove the wood shingles from the simulated roof, and then apply wood shakes. Point out any differences between the procedures.

8. Explain that wood shakes are also available in 8′ panels. Discuss the preparation necessary for these panels, and the proper procedure for applying them.

REROOFING OVER OLD SHINGLES

1. Discuss the preparation necessary for reroofing over old shingles.

2. Describe the application of new mineral fiber shingles.

ROOFING TILE

1. Identify the types of roofing tile that are commonly used. Have students list advantages of using roofing tile over other types of roofing products.
2. Discuss the weight of tile and the need for strong roof framing to support the weight.

APPLICATION OF TILE

1. Explain the underlayment requirements for installing tile roofs as they relate to roof pitch.
2. Explain what battens are and their use in the installation of tile.
3. Discuss lugs and their purpose.
4. Compare valley, hip, and ridge installations of tile with other types of roofing material.
5. Explain the term "sweat sheet" and discuss its proper application.
6. Discuss how cuts may be made on tile.
7. Identify materials required for installation at hips, ridges, and rakes. Discuss methods of installation.
8. Go over installation of tile in areas with severe weather conditions or on tall structures.

SHEET METAL ROOFING

1. Emphasize the need to use either galvanized sheets or those coated with aluminum-zinc alloy for permanent structures. Explain that sheets with thinner coatings will likely need to be painted every few years.
2. Discuss the proper means of lapping galvanized sheets to ensure watertightness. Stress that sheets should overlap a minimum of one-and-one-half corrugations for sufficient effectiveness.
3. Obtain a few lead-headed nails and galvanized nails with lead washers. Discuss the importance of using the correct type of nail for this type of roofing.

ALUMINUM ROOFING

1. Identify areas of the country where aluminum roofing should not be used. Explain the effects of salt-laden spray on aluminum.
2. Stress the importance of not allowing aluminum roofing material to come into contact with other types of metals. Describe what should be done if there are places where this contact cannot be avoided.
3. Identify the type of preparation that is necessary for aluminum roofing. Discuss the type of nails that should be used to install aluminum roofing.

TERNE METAL ROOFING

1. Describe the composition and manufacture of Terne metal roofing.
2. Discuss the grades and sizes of Terne metal roofing that are available.

GUTTERS

1. Describe the relationship between gutters or eaves troughs and downspouts. Explain the difference between a gutter and eaves trough.
2. Have students list the types of wood commonly used for wood gutters, and have them explain why they think these are good materials to use.
3. Using Reproducible Master 11-5, *Metal Gutter System,* identify the parts of such a system.
4. Discuss how sizes of gutters are determined for different sizes of roofs.
5. Emphasize that metal gutters should be sloped approximately 1″ for every 12′ to 16′ of length for proper drainage.

ESTIMATING MATERIALS

1. Discuss how the amount of roofing materials for a structure can be determined from the figures obtained in estimating the sheathing.
2. Describe an alternative method of estimating the roof area by determining the total ground area of the structure and then multiplying by a percentage factor of the roof slope.
3. Explain how the number of squares of roofing material can be obtained from the total square footage of roof surface.

UNIT REVIEW

1. Review the unit objectives. Be sure that the students fully understand each objective.
2. Assign *Important Terms, Test Your Knowledge* questions, and *Outside Assignments* on pages 314-316 of the text. Review the answers in class.
3. Assign pages 73-80 of the workbook. Review the answers in class.

EVALUATION

1. Use Procedure Checklist—*Installing Asphalt Shingles,* to evaluate the techniques that students used to perform these tasks.
2. Use Procedure Checklist—*Applying Wood Shingles,* to evaluate the techniques that students used to perform these tasks.
3. Use Unit 11 Quiz for in-class evaluation. Correct the quizzes and return them to the students for review.

ANSWERS TO TEST YOUR KNOWLEDGE
TEXT PAGE 315

1. slope (or pitch)
2. top lap
3. 36″
4. 36″
5. 15
6. 1/8
7. four
8. A. install staples, generally speaking, 5/8″ below adhesive strip
   B. adjust staple gun pressure for proper staple application
   C. repair the hole with asphalt cement if a staple must be removed
9. 336.5 mm × 984 mm
10. base
11. 2 in 12
12. See page 284 and Figure 11-14. Auxiliary roof deck on high side of chimney to divert flow of water and prevent ice and snow buildup behind the chimney.
13. See page 292. Slag, gravel, crushed stone, or marble chips.
14. gravel stop
15. perpendicular
16. 1/4
17. 1 1/2
18. cedar shakes
19. He or she tells the supplier the type of roofing materials; the supplier obtains truss rafters designed by engineers to carry the extra weight. On reroofing jobs, added bracing must be given stick-built rafters.
20. Yes. Generally, two plies of type 15 asphalt underlayment are laid down with hot mopping between layers. An alternative, though expensive, is to use a membrane or ice and water barrier.
21. 1 1/2
22. lead, tin
23. Snow accumulations slide off more easily.
24. conductors or downspouts

ANSWERS TO WORKBOOK QUESTIONS
PAGES 73-80

1. color
2. fascia
3. square
4. A. Side lap.
   B. Head lap.
   C. Exposure.

5. A. 4″
   B. 2″
   C. Metal drip edge.
   D. Underlayment.
6. B. 2′
7. A. 18″
   B. 36″
   C. 12″
8. 1/8
7. 3/8″
9. A. step-flashing
10. A. Annular thread.
    B. Plain barbed.
    C. Spiral thread.
11. A. 7/8″
    B. 1″
    C. 1 1/4″ or 1 1/2″
    D. 1 1/4″ or 1 1/2″
12. A. 12″
    B. 36″
    C. 5 5/8″
13. B. a strip from which one-half tab has been cut away
14. cap flashing
15. A. 1″
    B. 5 1/2″
16. B. 24″
17. A. 1″
18. A. 18″
19. D. 1 3/4″
20. A. Gravel stop.
    B. Built-up roofing.
    C. Sheathing.
    D. Soffit.
    E. Fascia.
21. A. Cant strip.
    B. 8″
    C. Metal flashing.
22. 24″
25 sq. ft.
23. B. 3
24. A. 1 1/2″
    B. 1″ to 2″
    C. 1/4″ to 3/4″
    D. 1/4″
25. A. 4 to 12
26. A. Straight split.
    B. Handsplit and resawn.
    C. Taper split.
27. 30 lb. roofing felt

28. A. 5.8
    B. 10.25
29. A. battens
    B. 24
    C. lugs
30. C. 1 1/2
31. A. lead and tin
32. A. Gutter.
    B. Downspout/conductor.
    C. Outside miter.
    D. Outlet tube.
    E. Elbow.
    F. Fascia hanger.
    G. Strap hanger.
    H. Pipe band.
33. 23
34. 37
35. 86

36. 22
37. pitch: 1/2
slope: 12

ANSWERS TO UNIT 11 QUIZ

1. False.
2. True.
3. False.
4. False.
5. D. All of the above.
6. B. drip edge
7. B. 24″
8. D. All of the above.
9. E
10. B
11. A
12. D
13. C

Modern Carpentry Instructor's Manual

Roofing Materials and Methods

Reproducible Master 11-1 (Part A)
Shingling Valleys

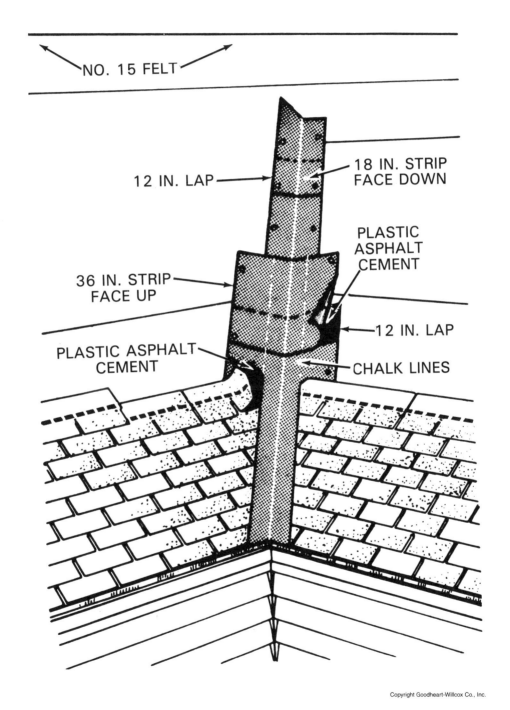

NO. 15 FELT

12 IN. LAP

18 IN. STRIP
FACE DOWN

PLASTIC
ASPHALT
CEMENT

36 IN. STRIP
FACE UP

12 IN. LAP

PLASTIC ASPHALT
CEMENT

CHALK LINES

Reproducible Master 11-1 (Part B)
Shingling Valleys

36 IN. ROLL ROOFING
50 LB. OR HEAVIER

EXTRA NAIL IN
END OF STRIP

6 IN.
MIN.

EACH STRIP TO
EXTEND AT
LEAST 12 IN.
BEYOND CENTER
OF VALLEY

Reproducible Master 11-1 (Part C)
Shingling Valleys

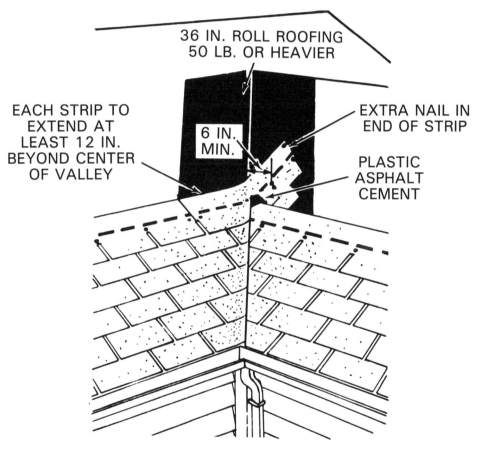

36 IN. ROLL ROOFING
50 LB. OR HEAVIER

EACH STRIP TO
EXTEND AT
LEAST 12 IN.
BEYOND CENTER
OF VALLEY

6 IN.
MIN.

EXTRA NAIL IN
END OF STRIP

PLASTIC
ASPHALT
CEMENT

Reproducible Master 11-2
Six Inch Shingling Method

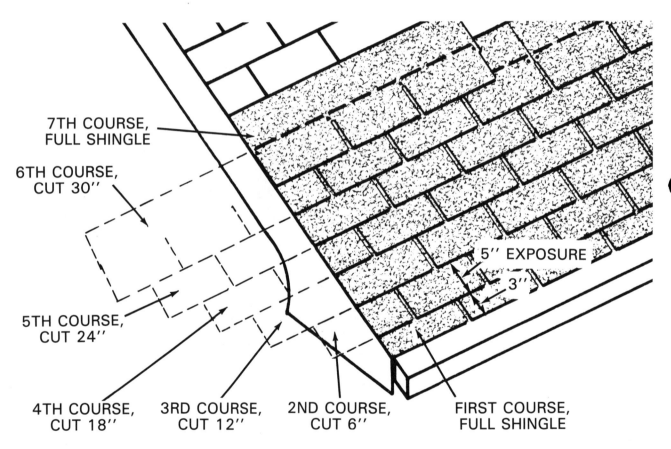

7TH COURSE, FULL SHINGLE

6TH COURSE, CUT 30"

5TH COURSE, CUT 24"

4TH COURSE, CUT 18"

3RD COURSE, CUT 12"

2ND COURSE, CUT 6"

FIRST COURSE, FULL SHINGLE

5" EXPOSURE

3"

Reproducible Master 11-3

Four Inch Shingling Method

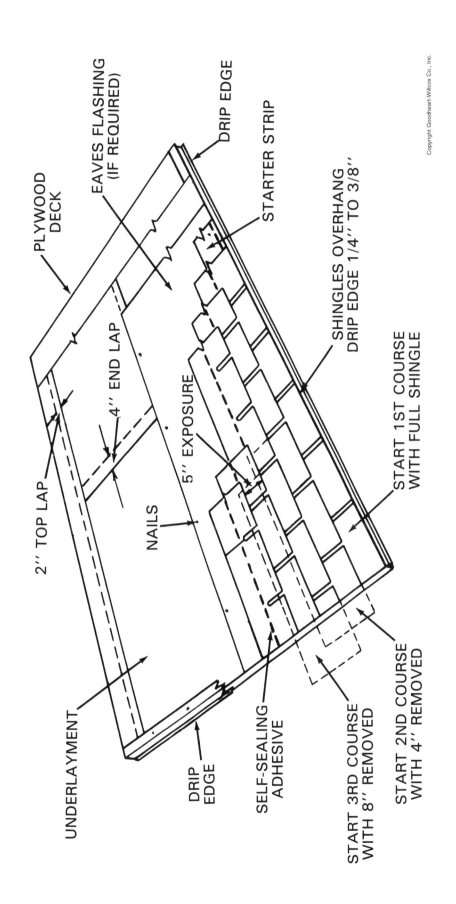

PLYWOOD DECK

EAVES FLASHING (IF REQUIRED)

DRIP EDGE

STARTER STRIP

SHINGLES OVERHANG DRIP EDGE 1/4'' TO 3/8''

2'' TOP LAP

4'' END LAP

5'' EXPOSURE

NAILS

START 1ST COURSE WITH FULL SHINGLE

UNDERLAYMENT

DRIP EDGE

SELF-SEALING ADHESIVE

START 3RD COURSE WITH 8'' REMOVED

START 2ND COURSE WITH 4'' REMOVED

Reproducible Master 11-4
Vent Stack Flashing

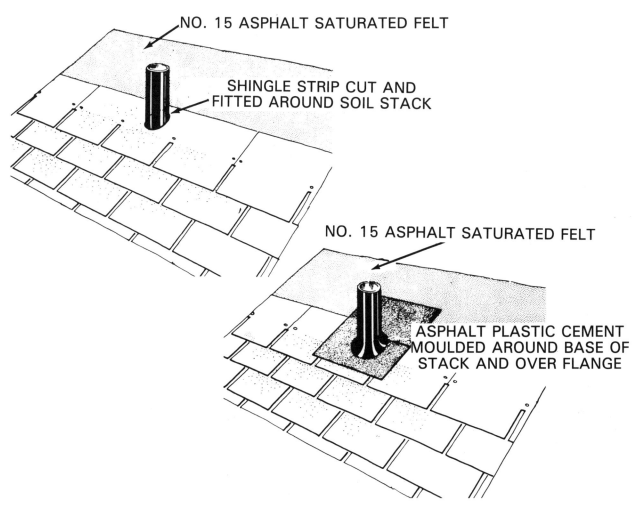

NO. 15 ASPHALT SATURATED FELT

SHINGLE STRIP CUT AND FITTED AROUND SOIL STACK

NO. 15 ASPHALT SATURATED FELT

ASPHALT PLASTIC CEMENT MOULDED AROUND BASE OF STACK AND OVER FLANGE

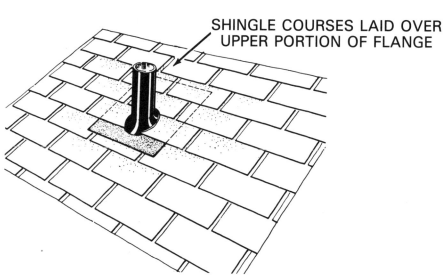

SHINGLE COURSES LAID OVER UPPER PORTION OF FLANGE

Reproducible Master 11-5
Metal Gutter System

KEY	DESCRIPTION	KEY	DESCRIPTION
1	5" K GUTTER	9	K OUTLET TUBE (With Flange)
2	3" SQUARE CORRUGATED DOWNSPOUT	10	5" K FASCIA HANGER
3	5" K MITER (Outside)	11	5" K STRAP HANGER
4	5" K MITER (Inside)	12	7" SPIKE (Aluminum) 5" FERRULE (Aluminum)
5	5" K SLIP JOINT CONNECTOR	13	5" K STRAINER
6	5" K END CAP LEFT OR RIGHT	14	3" PIPE BAND (Ornamental)
7	5" x 3" K END SECTION WITH OUTLET TUBE	15	TOUCH-UP PAINT SPRAY ON TOUCH-UP PAINT (White Only)
8	3" SQUARE CORRUGATED 75° ELBOW OR 60° ELBOW STYLE A AND B	16	GUTTER SEAL (Tube or Cartridge)

Unit 11 Procedure Checklist
Applying Wood Shingles

Name _____ Total _____

❑ Installs proper type of sheathing to roof deck, making sure it is adequate to hold the roofing materials and fasteners; cleans off any chips or other scrap material.	5	4	3	2	1	Neglects to prepare roof deck; does not clean off roof prior to installing underlayment.
❑ Properly applies underlayment using proper top and side laps.	5	4	3	2	1	Neglects to install underlayment.
❑ In cold climates, installs flashing at eaves; installs flashing at valleys (if necessary); installs flashing at vertical wall lines (if necessary).	5	4	3	2	1	Neglects to install any flashing.
❑ Selects the correct type and length of nail for the roofing material being used.	5	4	3	2	1	Selects the wrong type of nail for roofing.
❑ Doubles or triples first course of shingles; uses appropriate number of nails per shingle with the correct nailing pattern; starts second course offset with joints in first course; uses a straight edge to align rows of shingles; places pieces of beveled siding along gable ends to prevent water from dripping off edge.	5	4	3	2	1	Neglects to use doubled or tripled shingles for first course; neglects to follow nailing pattern; does not offset second course of shingles; "eyeballs" straightness of rows; neglects to use beveled siding along edge.
❑ Installs wood cap shingles along hips and ridges, making sure not to expose nails.	5	4	3	2	1	Neglects to install hip or ridge shingles; if cap shingles were installed, no attention was paid to concealing nail heads.

Unit 11 Procedure Checklist
Installing Asphalt Shingles

Name _____ Total _____

☐ Properly prepares the roof deck, making sure it is adequate to hold the roofing materials and fasteners; repairs any large knot holes with sheet metal; cleans off any chips or other scrap material.	5	4	3	2	1	Neglects to prepare roof deck, forgets to cover large knot holes; does not clean off roof prior to installing underlayment.
☐ Properly applies underlayment using proper top and side laps.	5	4	3	2	1	Neglects to install underlayment.
☐ Determines the correct shape of drip edge; accurately cuts the drip edge to length; properly positions drip edge—over the underlayment at the rake and under the underlayment along the eaves.	5	4	3	2	1	Must be told the correct length of drip edge; does not measure length of drip edge correctly resulting in poor fit; does not properly install drip edge.
☐ In cold climates, installs flashing at eaves; installs flashing at valleys (if necessary); installs flashing at vertical wall lines (if necessary).	5	4	3	2	1	Neglects to install any flashing.
☐ Determines best means for installing shingles; snaps chalk lines; carefully places shingles at locations along the roof.	5	4	3	2	1	Neglects to snap chalk lines; stacks all shingles in one pile, resulting in more work later.
☐ Selects the correct type and length of nail for the roofing material being used; properly installs starter strip.	5	4	3	2	1	Selects the wrong type of nail for roofing; forgets to install starter strip.

181

Unit 11 Procedure Checklist (continued)

❏ Starts first course with full shingle; uses appropriate number of nails per shingle with the correct nailing pattern; cuts and starts second course with a cut strip; follows a uniform laying pattern.

5 4 3 2 1

Uses cut shingle for first course; neglects to follow nailing pattern; lays shingles in irregular pattern.

❏ Flashes around chimney (if necessary); constructs chimney saddle (if necessary); flashes vent stack projections; carefully installs shingles around these obstructions.

5 4 3 2 1

Neglects to install flashing around roof obstructions; carelessly installs shingles around these obstructions.

❏ Determines whether to use special hip or ridge shingle or to make them from standard shingles; properly installs hip and ridge shingles.

5 4 3 2 1

Neglects to install hip or ridge shingles.

Unit 11 Quiz
Roofing Materials

Name _____ Score _____

True-False

Circle T if the answer is True or F if the answer is False.

T   F   1. One square of asphalt shingles covers 150 square feet.

T   F   2. Saturated felts are used for sheathing paper under shingles.

T   F   3. One ply is generally sufficient for a built-up roof on a residential structure.

T   F   4. When driving nails into wood shingles, the heads should be driven about 3/16 inch below the surface.

Multiple Choice

Choose the answer that correctly completes the statement. Write the corresponding letter in the space provided.

_____   5. Materials commonly used for pitched roofs include _____.

A. asphalt shingles
B. slate and tile
C. roll roofing
D. All of the above.

_____   6. The _____ is bent downward over the edge of the roof, causing rainwater to drip free of the cornice construction.

A. flashing
B. drip edge
C. underlayment
D. starter strip

_____   7. It is recommended that ice and water barrier be installed at the eaves and extending _____ inside the walls.

A. 12″
B. 24″
C. 30″
D. 36″

_____   8. Wood shingles are made from _____.

A. cypress
B. western red cedar
C. redwood
D. All of the above.

Identification

Identify recommendations for types of roofing according to pitch.

_____ 9. Square tab strip shingles (low slope application).

_____ 10. All styles of shingles.

_____ 11. Slope.

_____ 12. Roll roofing (concealed nails).

_____ 13. Roll roofing (exposed nails).

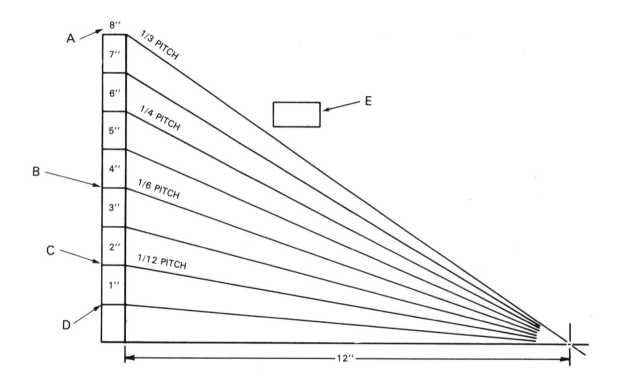

12

Windows and Exterior Doors

OBJECTIVES

Students will be able to:

- ❏ Discuss standards for window and door fabrication.
- ❏ Recognize the types of windows.
- ❏ Calculate required rough openings and interpret a "window schedule."
- ❏ Explain how window frames are adjusted for wall thickness.
- ❏ Summarize procedures for installing a standard window.
- ❏ Describe procedures for installing a replacement window.
- ❏ Prepare a rough opening for the installation of a door frame.
- ❏ Describe the procedure for sliding glass door installation.
- ❏ Explain the correct construction of garage door frames.
- ❏ Select appropriate garage door hardware.
- ❏ Describe the proper procedure for installing one type of modern bow or box bay window unit.

INSTRUCTIONAL MATERIALS

Text: Pages 317-350
　　Important Terms, page 348
　　Test Your Knowledge, page 348
　　Outside Assignments, page 350
Workbook: Pages 81-88
Instructor's Manual:
　　Reproducible Master 12-1, *Window Framing Detail*
　　Reproducible Master 12-2, *Plumbing Door Jambs*
　　Procedure Checklist—*Installing Movable Windows*
　　Procedure Checklist—*Installing Doorframes*
　　Unit 12 Quiz

TRADE-RELATED MATH

Glass blocks are commonly used for interior decorative work, as well as exterior windows. The widths and heights of the openings can be determined by multiplying the number of units, horizontally and vertically, multiplying each by the nominal block size, and then adding 3/8″. For example to determine the size opening required for a panel of 6″ blocks that is to be 5 units wide × 8 units high, the following calculations would be used:

$$\text{width} = 5 \times 6 + \frac{3}{8} = 30\,\frac{3}{8}'' = 2' - 6\,\frac{3}{8}''$$

$$\text{height} = 8 \times 6 + \frac{3}{8} = 48\,\frac{3}{8}'' = 4' - \frac{3}{8}''$$

INSTRUCTIONAL CONCEPTS AND STUDENT LEARNING EXPERIENCES

MANUFACTURE

1. Describe the manufacture of window and door units. Ask the students to list the materials commonly used for windows and doors. Discuss advantages and disadvantages of using these materials.

TYPES OF WINDOWS

1. Identify the three general categories of windows: sliding, swinging, and fixed. List the basic window designs that are included in each category.
2. Describe the basic components and operation of a double-hung window.
3. Describe the basic components and operation of a horizontal sliding window.
4. Describe the basic components and operation of a casement window. Define the term "crank operator" and discuss its use with a casement window.
5. Describe the basic components and operation of an awning window.

6. Describe the basic components and operation of a hopper window. Contrast its operation to that of an awning window. Discuss some of the inconveniences caused by hopper windows.
7. Have the students describe the purpose of a multiple-use window.
8. Describe the basic components and operation of a jalousie window. Explain why they find limited use in the northern climates.
9. Describe the basic components and operation of a fixed window.

WINDOW HEIGHTS
1. Refer to Figure 12-8. Have students note the different size windows that are used for different rooms of the house. Also have them note that the standard window height from the finished floor to the bottom side of the window head is 6'-8''. Explain why a standard height is generally used.

WINDOW GLASS
1. Discuss the manufacture of window glass. Have the students refer to Figure 12-10 for a list of standard thicknesses that are manufactured.

ENERGY EFFICIENT WINDOWS
1. Define the term "R value" and how it relates to windows. Stress the importance of using windows with high R values for energy conservation.
2. Have students state reasons why two or three layers of glass might be used in a window unit. Discuss the advantages of double and triple glazing.
3. Define "emissivity" and how it relates to windows. Describe how "low-e" windows help to conserve energy.

SCREENS
1. Discuss the purpose of a screen. Identify materials that are commonly used for screens.

MUNTINS
1. Describe the difference between muntins used in older construction and muntins used in modern construction. Ask students to identify the different patterns that are commonly used.

PARTS OF WINDOWS
1. Identify the following parts of a window: mullion, drip cap, head, sill, and jamb. Stress the importance of using the correct terminology when referring to windows so that they can communicate effectively with other tradesworkers.

WINDOWS IN PLANS AND SPECIFICATIONS
1. Discuss how windows and exterior doors are shown in working drawings. Emphasize that these items are generally dimensioned to the middle of the opening for frame construction and to the edge of the opening for masonry construction.
2. Have students identify the direction of swing using a stock plan for a house. Mention that sliding windows will probably be noted as such so as to distinguish them from fixed windows.

WINDOW SIZES
1. Explain that in addition to the type and position of the window, the carpenter should also know the glass size, sash size, rough frame opening, and masonry or unit opening. Given a stock plan, have the students identify these measurements. Stress that the horizontal measurement is always listed first.
2. Describe the items that are usually found in a window and door schedule. Explain how letters found on a plan for a structure will generally be cross referenced to a window and door schedule.

DETAILED DRAWINGS
1. Using manufacturer's literature or Figure 12-24, have the students note how different types of construction affect windows installation.

JAMB EXTENSIONS
1. Discuss how a jamb extension can be used to compensate for different wall thicknesses.

STORY POLE
1. Review the use of a story pole when installing doors and windows. Discuss the advantages of using a story pole. Mention that even though the heads of the windows and doors are at the same height, the bottoms will vary.
2. Using Reproducible Master 12-1, *Window Framing Detail,* have the students note the other dimensions that should be added to a story pole when checking the height of certain parts of the window frame.

INSTALLING WINDOWS
1. Describe the proper means of storing windows on a job site. Note that they should be allowed to acclimate to the humidity of the locality before being installed.
2. Discuss the preparation that should be made before installing a window unit, such as checking

the rough opening for the correct size and priming the window (if necessary).

3. Explain the proper procedure for installing a window. Stress the importance of properly leveling and plumbing the window unit before permanently fastening it into position.

4. Discuss steps that might be taken after the installation of the window units to protect them against damage and dirt.

5. Using Figure 12-39, explain the installation of large fixed window units. Emphasize that larger units generally require extra precautions in handling and installation.

6. Refer to Figure 12-38. Stress the need for clearance between sealed insulating glass and its frame.

GLASS BLOCKS

1. Have the students list advantages of using glass block for interior and exterior wall applications.

2. List the three nominal sizes of glass blocks. Mention that even though a carpenter will usually not make the actual installation, it is helpful for them to have knowledge of the design requirements.

3. Discuss the procedure for installing small glass block panels. Refer to Figure 12-45 for illustrations of this procedure.

4. Compare the installation of a small glass block panel to that of a large glass block panel. Emphasize that wall tie and anchors must be used for larger panels to add stability to the panel.

5. Explain how to determine the opening required for a glass block installation.

REPLACING WINDOWS

1. Cite reasons for replacing older window units. Discuss the information needed to order new window units. Mention that some newer replacement window units are made to size, and thus made to fit the rough opening for existing windows.

2. Describe the procedure for removing the older units and installing new units.

3. Discuss reasons for installing skylights.

EXTERIOR DOOR FRAMES

1. Identify the sizes of exterior doors commonly used in residential construction. Note that rear and service doors are usually narrower than front entrance doors.

2. Have the students refer to the local building code

to determine the minimum exterior door width.

3. Discuss why the door head and jambs are made of 5/4″ stock.

4. Explain that exterior doors in residential construction swing inward and the rabbets for the head and jambs must be located on the inside.

5. Have the students list materials that are commonly used for sills.

INSTALLING DOOR FRAMES

1. Identify the preparation necessary before installing the door frame.

2. Describe the procedure for installing the door frame using Reproducible Master 12-2, *Plumbing Door Jambs*. Impress upon your students the reason why nails used to secure the frame should not be driven in all the way until a final check has been made.

INSTALLING PREHUNG DOOR UNITS

1. Discuss the advantages of using prehung door units.

2. Describe the procedure for installing prehung doors. Stress the importance of following the manufacturer's recommendations when installing a prehung door unit.

SLIDING GLASS DOORS

1. Have students list reasons why sliding glass doors have become so popular in residential construction.

2. Explain how sliding and fixed doors are indicated on working drawings and manufacturers' literature.

3. Describe the procedure used to install sliding glass doors. Stress that an "X" should be placed on the sliding glass doors and fixed doors with tape or washable paint to prevent anyone from walking into the doors.

GARAGE DOORS

1. Have the students list the three types of garage doors. Mention that roll-up doors are the most common type in residential construction.

GARAGE DOOR FRAMES

1. Identify the most common garage door widths and heights.

2. Discuss the procedure for installing garage doors. Emphasize that the frame construction is very similar to frame construction for regular door frames.

3. Discuss the purpose of garage door hardware and counterbalances.

BOW AND BOX BAY WINDOWS

1. Discuss differences between "stick-built" bay windows and the prefabricated units known as "bow" and "box bay windows."
2. Explain the procedure for installing a typical prefabricated unit.

UNIT REVIEW

1. Review the unit objectives. Be sure that the students fully understand each objective.
2. Assign *Important Terms, Test Your Knowledge* questions, and *Outside Assignments* on pages 348-350 of the text. Review the answers in class.
3. Assign pages 81-88 of the workbook. Review the answers in class.

EVALUATION

1. Use Procedure Checklist—*Installing Movable Windows*, to evaluate the techniques that students used to perform these tasks.
2. Use Procedure Checklist—*Installing Doorframes*, to evaluate the techniques that students used to perform these tasks.
3. Use Unit 12 Quiz for in-class evaluation. Correct the quizzes and return them to the students for review.

ANSWERS TO TEST YOUR KNOWLEDGE TEXT PAGE 348

1. millwork
2. ponderosa pine
3. casement window
4. 6'-8"
5. 0.88, 2.00
6. jamb
7. drip cap
8. 3'-5"
9. jamb extensions
10. level, sill
11. 1/4
12. 3 7/8
13. nominal
14. lintel
15. 16
16. roll-up
17. 16
18. torsion
19. wood bracing, cable suspension

ANSWERS TO WORKBOOK QUESTIONS PAGES 81-88

1. millwork
2. C. National Woodwork Manufacturers
3. double-hung
4. A. Fixed unit.
   B. Operating unit.
   C. Operator.
   D. Latch.
   E. Mullion.
5. awning
6. B. 6'-8"
7. B. 2.00
8. C. floating
9. B. 3/16"
10. A. Double-hung.
    B. Casement.
    C. Hopper.
    D. Awning.
11. C. 3'-4"
12. B. edge
13. A. Side jamb.
    B. Sill.
    C. Head.
14. mullion
15. B. 3/4"
16. B. jamb extension
17. C. raise frame to correct height as marked on story pole
18. B. 16" O.C.
19. A. Glazing clip.
    B. Glazing sealant.
    C. Setting block.
    D. 3/16"
    E. 1/4"
20. A. 3 7/8 × 11 3/4 × 11 3/4
21. Width: 4'-6 3/8"
    Height: 5'-6 3/8"
22. B. remove inside stops and lift out lower sash
23. C. 3 to 12
24. A. coffered
25. A. 6'-8"
    B. 2'-6"
26. A. Head.
    B. Sill.
    C. Jamb.
27. D. 2 1/2"
28. hinges
29. C. 16d
30. A. Screen.
    B. Sill support.

C. Operating panel.

D. Threshold.

31. B. a bead of sealing compound be applied

32. swing-up

33. C. 7'-0"

34. torsion

35. Clear outer pane.

An air space.

A special coating on the air-gap side of the inner pane.

ANSWERS TO UNIT 12 QUIZ

1. False.

2. False.

3. True.

4. True.

5. D. Both B and C.

6. B. 6'-8"

7. A. hopper

8. B. 1/2"

9. B

10. E

11. C

12. G

13. F

14. A

15. D

Reproducible Master 12-1
Window Framing Detail

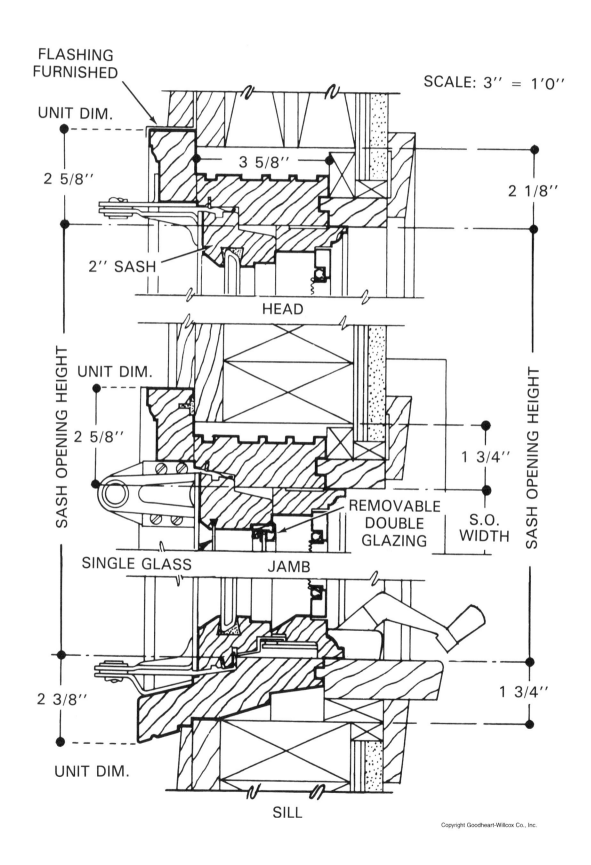

FLASHING
FURNISHED

UNIT DIM.

SCALE: 3'' = 1'0''

2 5/8''

3 5/8''

2 1/8''

2'' SASH

HEAD

SASH OPENING HEIGHT

UNIT DIM.

2 5/8''

1 3/4''

REMOVABLE
DOUBLE
GLAZING

S.O.
WIDTH

SASH OPENING HEIGHT

SINGLE GLASS

JAMB

2 3/8''

1 3/4''

UNIT DIM.

SILL

Reproducible Master 12-2
Plumbing Door Jambs

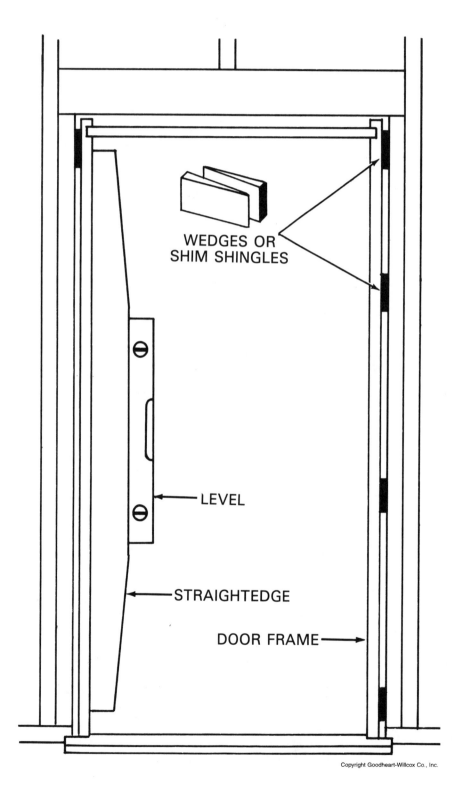

WEDGES OR
SHIM SHINGLES

LEVEL

STRAIGHTEDGE

DOOR FRAME

Unit 12 Procedure Checklist
Installing Movable Windows

Name _____ Total _____

❏ Moves all windows to vicinity of installation; unpacks and checks for damage (does not remove brace); checks plumb and level of rough openings; checks size of rough opening.	5	4	3	2	1	Moves only one window to needed location, then begins installing it; neglects to check for damage (removes bracing); forgets to check size of rough opening.
❏ Refers to manufacturer's recommendations and primes windows (if necessary).	5	4	3	2	1	Primes windows when it is not needed, or neglects to prime windows if it is needed.
❏ Installs window from outside; temporarily secures window in opening.	5	4	3	2	1	Tries to install window from inside; neglects to secure window in opening.
❏ Places wedge blocks under sill and raises it to proper height; levels window; plumbs the side jambs; checks squareness in window corners.	5	4	3	2	1	Rests window on sill (does not raise it to proper height); neglects to plumb or level window; neglects to square window.
❏ Drives nails into bottom, then top of side casing; levels and plumbs again; checks window to make sure it works properly.	5	4	3	2	1	Nails into top and bottom casing (possibly breaking window); neglects to plumb or level window; does not check window to make sure it works properly.
❏ Nails window permanently into place using appropriate types of nails; uses nail set to drive nail head below surface; covers window with plastic if interior wall surface materials are to be applied.	5	4	3	2	1	Forgets to permanently nail window into position; does not use proper type of nail; neglects to cover window with plastic.

Unit 12 Procedure Checklist
Installing Doorframes

Name _____ Total _____

❏ Checks size of rough opening; cut out sill area (if necessary); install flashing (if necessary).	5	4	3	2	1	Neglects to check size of rough opening; does not make adjustments to allow for larger door; forgets to install flashing if required.
❏ Places frame in opening, centering it horizontally; secures door frame with temporary brace; levels the sill using wedges and blocking; makes sure the sill is well-supported.	5	4	3	2	1	Pushes frame to one side of opening; neglects to support frame with brace; forgets to level sill; does not provide adequate support for sill.
❏ Drives nail through casing and into wall frame; inserts wedges or blocking between studs and jambs; adjusts wedge until frame is plumb.	5	4	3	2	1	Neglects to drive nails into casing (or misses wall frame); does not plumb frame.
❏ Places additional wedges and blocks between jamb and studs; drives nails through the jamb wedge and into the stud.	5	4	3	2	1	Does not support frame with additional wedges and blocks.
❏ Nails casing in place; installs piece of plywood over sill to protect it during further construction work.	5	4	3	2	1	Neglects to place casing in position; forgets to place plywood over sill.

Unit 12 Quiz

Windows and Exterior Doors

Name _____ Score _____

True-False

Circle T if the answer is True or F if the answer is False.

T   F   1. A building material with a high R-value has a low resistance to heat passage.

T   F   2. When specifying window sizes, the vertical dimension is always listed first.

T   F   3. Glass blocks have good insulating qualities.

T   F   4. Emissivity is the relative ability of a material to absorb or re-radiate heat.

Multiple Choice

Choose the answer that correctly completes the statement. Write the corresponding letter in the space provided.

_____   5. A(n) _____ window is classified as a swinging window.

   A. double hung
   B. casement
   C. awning
   D. Both B and C.

_____   6. In residential construction, the standard height from the bottom side of the window head to the finished floor is _____.

   A. 6'-4"
   B. 6'-8"
   C. 7'-0"
   D. 7'-6"

_____   7. A(n) _____ window is hinged along the bottom and swings inward.

   A. hopper
   B. awning
   C. casement
   D. jalousie

_____   8. The rough opening for a window should allow at least _____ clearance on the sides and 3/4" clearance above the head.

   A. 1/4"
   B. 1/2"
   C. 3/4"
   D. 1"

Identification

Identify the parts of the double glazed window unit.

_____ 9. Metal spacer.

_____ 10. Glass.

_____ 11. Primary seal.

_____ 12. Secondary seal.

_____ 13. Drying agent.

_____ 14. Corner clip.

_____ 15. Airspace.

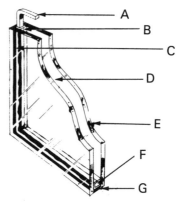

13

Exterior Wall Finish

OBJECTIVES

Students will be able to:
- ❏ Identify the parts of a cornice and rake.
- ❏ Describe cornice and rake construction.
- ❏ Illustrate approved methods of flashing installation.
- ❏ Describe how wood siding and shingles are applied.
- ❏ Estimate the amount of siding or shingles required for a specific structure.
- ❏ Discuss the proper application of bevel siding.
- ❏ List the most common siding choices and their characteristics.
- ❏ Discuss Exterior Insulation Finish Systems and their application.
- ❏ Demonstrate installation techniques for various siding materials.

INSTRUCTIONAL MATERIALS

Text: Pages 351-389
Important Terms, page 383
Test Your Knowledge, page 385
Outside Assignments, page 387
Workbook: Pages 89-96
Instructor's Manual:
Reproducible Master 13-1, *Typical Cornice Details,* Parts A and B
Reproducible Master 13-2, *Siding Application, Parts A* and *B*
Reproducible Master 13-3, *Typical Exterior Insulation Finish System*
Reproducible Master 13-4, *Brick Veneer Construction*
Procedure Checklist—*Installing Horizontal Wood Siding*
Unit 13 Quiz
Section III Exam

TRADE-RELATED MATH

When determining the amount of horizontal siding required for a structure, the net wall surface must be multiplied by a factor. The factor is used to compensate for cutting of joints, overlap (in bevel siding), and other considerations. The factors are shown in Figure 13-28 of the text.

First, determine the net wall surface area by subtracting the areas of the openings from the gross wall area (height × perimeter). For example, if the wall height is 8′, the perimeter of the house is 140′, and door and window opening amount to 210 square feet, determine the net wall area.

Net area = gross area – area of openings
= (8 × 140) – 210
= 1120 – 210
Net area = 910 square feet

Next, multiply the net area by the appropriate factor from Figure 13-28. If 1 × 6 rustic shiplapped siding is to be used, the factor is 1.19. Multiply the factor by the net wall area.

Siding needed = net area × factor
= 910 × 1.19
Siding needed = 1083 square feet

INSTRUCTIONAL CONCEPTS AND STUDENT LEARNING EXPERIENCES

CORNICE DESIGN

1. Using Parts A and B of Reproducible Master 13-1, *Typical Cornice Details,* discuss the various cornice designs. Mention that most modern residential structures have boxed cornices.
2. Define the following terms associated with a boxed cornice, making sure that the students understand their relationship to the others: fascia board, ledger strip, frieze, plancier (soffit), and

lookout. Have the students identify the types of material(s) that are used for each of these parts.

3. Discuss the properties that the exterior members should have, including good painting and weathering characteristics, easy workability, freedom from warp, and decay resistance.

CORNICE AND RAKE CONSTRUCTION

1. Describe the procedure used to construct a cornice with a horizontal soffit. Emphasize that since this is finish work, that joints should be mitered to hide unsightly end grain. Mention that rust-resistant nails or screws should be used to secure the soffit into position. If regular casing or finish nails are used, they must be countersunk and the holes filled with putty.
2. Discuss the procedure used to construct the rake so that it matches the cornice.
3. Discuss the advantages of using prefabricated cornice materials.
4. Describe the purpose of screened vents in a soffit.
5. Explain the procedure used to construct a cornice using prefabricated cornice materials.
6. Identify the three basic components of soffit systems made of prefinished metal panels.
7. Discuss the procedure for constructing and hanging metal soffits.

WALL FINISH

1. Explain that the upper level of finish material is usually applied first so that scaffolding can be attached directly to the wall frame along the lower section.
2. Identify the five most common types of horizontal siding and discuss their sizes. Mention that bevel siding is most commonly used. Discuss advantages and disadvantages of all the types.
3. Describe the steps that should be taken to protect the siding when it is delivered to the construction site. Once again, stress that since this is finish materials, it should be handled carefully.
4. Introduce stucco and Exterior Insulation Finish Systems as alternatives to wood, aluminum, or vinyl siding.

WALL SHEATHING AND FLASHING

1. Discuss preparations that should be made prior to applying siding. Mention that, depending upon the type of sheathing, sheathing (building) paper may or may not be required. Stress that coated papers or laminated papers should not be used

because they have a high moisture resistance and will act as a vapor barrier.
2. Using part A of Reproducible Master 13-2, *Siding Application*, illustrate how flashing should be used over drip caps of doors and windows.
3. Discuss the minimum lap that should be used for horizontal siding.
4. Explain the procedure for laying out the position of the siding using a story pole. Mention that a carpenter should try to adjust the courses of siding to come out even with the tops and bottoms of windows, if possible.
5. Describe the procedure for installing horizontal siding using part B of Reproducible Master 13-2, *Siding Application*. Emphasize that nails should be driven into the studs, and note their placement along the butt edge of the siding.
6. Discuss the techniques used at the inside and outside corners.

ESTIMATING AMOUNT OF SIDING

1. Discuss the method used to estimate the amount of siding required for a residential structure. Make sure to include the waste factor.

VERTICAL SIDING

1. Identify the type of materials commonly used for vertical siding.
2. Describe how vertical wood siding is attached to a structure. Discuss the type of preparation (if any) that might be required prior to installation.
3. Discuss the use of battens. Describe how the board and batten effects can be simulated without using vertical wood siding.

WOOD SHINGLES

1. Explain the recommendations for maximum exposure of 16″, 18″, and 24″ wood shingles.
2. Define "double coursing," and describe how it can used effectively as a wall covering. Describe how shingles are applied using the single-coursing method.
3. Discuss the difference between applying shingles for a roof and a side wall. Describe the application of wood shingles using the single-coursing method.
4. Explain how quantities of wood shingles are estimated for side walls.

SHINGLE AND SHAKE PANELS

1. Discuss the advantages of using shingle or shake panels over using individual shingles or shakes.

RE-SIDING WITH WOOD SHINGLES

1. Describe the preparation necessary to install wood shingles over existing siding or other wall coverings.

MINERAL FIBER SIDING

1. Explain the recommendations for maximum exposure of mineral fiber siding.
2. Describe the procedure for applying mineral fiber siding. Explain that the type of material it is being installed over will dictate how the siding is applied.

PLYWOOD SIDING

1. Describe how plywood siding can be used to enhance the appearance of other exterior finish materials.
2. Have the students identify the types of wood that are commonly used for exterior plywood siding. Ask them to describe the types of finishes that may be used.
3. Identify the different thicknesses of plywood siding that are available, and describe situations where each type can be used.
4. Discuss the application of plywood siding to a residential structure. Emphasize that since plywood and hardboard provide tight, draft-free wall construction, that it is important to have an effective vapor barrier between the insulation and warm surface of the wall.
5. Refer to Figure 13-46. Discuss the different means of treating joints between plywood panels.
6. Have the students refer to the local building code to see if it includes any requirements regarding plywood or hardboard siding.

HARDBOARD SIDING

1. Compare installation of hardboard siding to that of plywood siding. Emphasize that some types of hardboard siding may expand more than plywood siding and thus should be considered when installing the panels.
2. Identify the common dimensions of hardboard siding.
3. Discuss the advantages of using a siding system over installing panels or individual shingles.

ALUMINUM SIDING

1. Describe the advantages of using aluminum siding.
2. Emphasize that the manufacturer's recommendations should be followed when installing aluminum siding.
3. Since aluminum is an electrical conductor, stress the possibility of an electrical hazard. Emphasize that the Aluminum Siding Association recommends connecting a No. 8 wire or larger from the aluminum siding to the electrical service ground or cold water service.

VINYL SIDING

1. Discuss the type of backing that should be used beneath vinyl siding.
2. Refer to Figure 13-57 for an illustration showing the types of accessories commonly used with vinyl siding.

STUCCO

1. Have students explain why balloon framing is the preferred framing method when applying a stucco finish.
2. Identify the types of bases used for stucco.
3. Briefly explain the procedure used to apply stucco.

EXTERNAL INSULATION FINISH SYSTEM (EIFS)

1. Using Reproducible Master 13-3, *Typical Exterior Insulation Finish System,* introduce this siding system and discuss its components.
2. Explain the EIFS insulating qualities.
3. Explain steps for applying EIFS siding materials. Refer to Figure 13-71.

BRICK OR STONE VENEER

1. Using Reproducible Master 13-4, *Brick Veneer Construction,* have the students note features of brick veneer construction including the base flashing, weep holes, and air space between the plywood sheathing and brick veneer. Have the students explain why these features are incorporated into brick veneer construction.

BLINDS AND SHUTTERS

1. Describe the historical significance of blinds and shutters along the sides of windows and doors. Discuss their purpose in this day and age.

UNIT REVIEW

1. Review the unit objectives. Be sure that the students fully understand each objective.
2. Assign *Important Terms, Test Your Knowledge* questions, and *Outside Assignments* on pages

383-387 of the text. Review the answers in class.

3. Assign pages 89-96 of the workbook. Review the answers in class.

EVALUATION

1. Use Procedure Checklist—*Installing Horizontal Wood Siding,* to evaluate the techniques that students used to perform these tasks.

2. Use Unit 13 Quiz for in-class evaluation. Correct the quizzes and return them to the students for review.

3. Use Section 3 Exam to evaluate the students' knowledge of information presented in Units 11-13.

ANSWERS TO TEST YOUR KNOWLEDGE
TEXT PAGE 385

1. ledger strip or frieze board (either)
2. lookouts
3. drop
4. edge-grain
5. one-fourth
6. 1 1/2
7. before
8. batten
9. 24
10. one-half
11. Allow carpenter to nail siding in the groove every 16″ or 24″, depending on spacing of studs.
12. 3/8
13. warm
14. 7/16
15. grounded
16. *Any four of the following:*
    Minimal maintenance costs.
    Minimal work to maintain.
    Has the appearance of fine wood siding or other siding materials.
    Waterproof interlocking joints.
    Tough and durable.
    Lightweight.
    Easy to install.
17. To provide additional insulation and provide a smooth surface for re-siding.
18. Nail loose siding; replace rotten boards.
    Remove old caulk around door and window trim.
    Remove downspouts, lighting fixtures, and moldings.
    Tie back shrubbery and tree branches close enough to be damaged or interfere with work.
19. False.
20. Rigid insulation board, base adhesive coat,

fiberglass mesh, finish coat of synthetic material.
21. Trowel, mason's level, mason's rule or steel tape, jointing tools, brick hammer, chisels, mason's line, power driven masonry saw, and joint rakers.
22. weep holes

ANSWERS TO WORKBOOK QUESTIONS
PAGES 89-96

1. A. doors
   B. windows
2. rafters
3. A. Fascia.
   B. Plancier/soffit.
   C. Vent.
   D. Lookout.
   E. Ledger.
   F. Frieze.
4. nailing strip
5. plancier/soffit
6. A. Lookout.
   B. Rake soffit.
   C. Fascia.
   D. Trim molding.
7. B. Secures sheathing to wall studs.
8. A. Clapboards.
   B. Bevel.
   C. Rabbeted-bevel.
   D. Rustic.
   E. Drop.
   F. Log Cabin.
9. A. Channel (or hanger).
   B. Soffit.
   C. Fascia.
   D. Channel (or hanger, or F channel).
10. A. 7 1/4″
11. D. 24″
12. B. Rigid polystyrene insulating board.
    C. Housewrap.
13. A. Flashing.
    B. Drip cap.
14. B. 1 1/2″
15. story pole
16. metal corner
17. before
18. above
19. 280
20. 1597
21. D. 7 1/2″
22. A. the center
23. Insulated
24. False.

25. B. 6″ O.C.
26. B. temperature
27. C. 7/16″
28. Aluminum Siding Assoc.
29. A. No. 8
30. B. extrusion
31. B. 1/4″
32. A. Building paper.
    B. Sheathing.
    C. Flashing.
    D. Brick.
    E. Space.
33. D. 48″
34. B. Shutters are usually attached with special concealed hinges

ANSWERS TO UNIT 13 QUIZ

1. True.
2. False.
3. False.
4. True.
5. C. fascia board
6. A. 1 1/2″
7. C. 8′
8. B. Four
9. A. Cement board, plywood, or OSB
10. B. Adhesive
11. C. Expanded polystyrene insulation board
12. D. Reinforcing mesh
13. E. Base coat of mortar
14. F. Surface coating (acrylic polymer emulsion)

ANSWERS TO SECTION 3 EXAM

1. True.
2. False.
3. False.
4. True.
5. True.
6. True.
7. True.
8. True.
9. True.
10. False.
11. C. fascia
12. A. square
13. B. centerline
14. A. saddle
15. D. eaves trough
16. B. 3′-0″
17. D. All of the above.
18. C. 4
19. B. Sheet metal
20. C
21. B
22. D
23. E
24. A
25. C
26. A
27. B
28. E
29. D
30. Step
31. 4
32. 5.8
    10.25
33. floating
34. 6′-8″
35. Exterior Insulated Finishing System
36. F
37. D
38. A
39. E
40. B
41. C
42. C
43. E
44. D
45. A
46. B
47. F

Reproducible Master 13-1 (Part A)
Typical Cornice Details

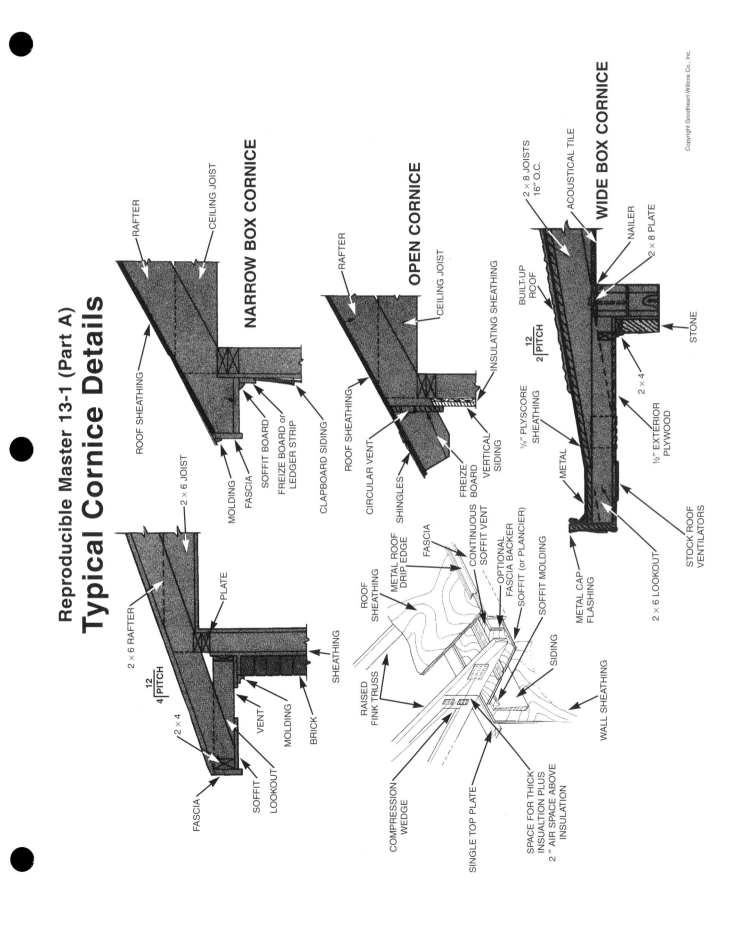

NARROW BOX CORNICE

- RAFTER
- CEILING JOIST
- ROOF SHEATHING
- 2 × 6 JOIST
- MOLDING
- FASCIA
- SOFFIT BOARD
- FREIZE BOARD or LEDGER STRIP
- CLAPBOARD SIDING

OPEN CORNICE

- RAFTER
- CEILING JOIST
- INSULATING SHEATHING
- ROOF SHEATHING
- CIRCULAR VENT
- SHINGLES
- FREIZE BOARD
- VERTICAL SIDING

WIDE BOX CORNICE

- 2 × 8 JOISTS 16" O.C.
- ACOUSTICAL TILE
- NAILER
- 2 × 8 PLATE
- STONE
- 2 × 4
- ½" EXTERIOR PLYWOOD
- BUILT-UP ROOF
- 12/2 PITCH
- ⅝" PLYSCORE SHEATHING
- METAL
- STOCK ROOF VENTILATORS
- 2 × 6 LOOKOUT
- METAL CAP FLASHING

- 2 × 6 RAFTER
- PLATE
- 12/4 PITCH
- 2 × 4
- VENT
- MOLDING
- BRICK
- SHEATHING
- FASCIA
- SOFFIT
- LOOKOUT

- ROOF SHEATHING
- METAL ROOF DRIP EDGE
- FASCIA
- CONTINUOUS SOFFIT VENT
- OPTIONAL FASCIA BACKER
- SOFFIT (or PLANCIER)
- SOFFIT MOLDING
- SIDING
- RAISED FINK TRUSS
- COMPRESSION WEDGE
- SINGLE TOP PLATE
- SPACE FOR THICK INSUALTION PLUS 2" AIR SPACE ABOVE INSULATION
- WALL SHEATHING

203

Reproducible Master 13-1 (Part B)
Typical Cornice Details

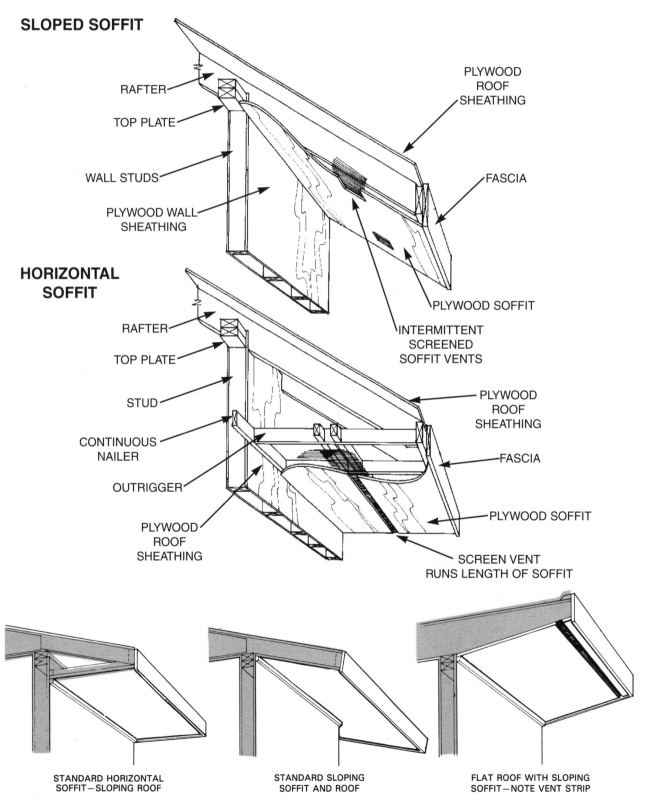

SLOPED SOFFIT

RAFTER

TOP PLATE

WALL STUDS

PLYWOOD WALL SHEATHING

PLYWOOD ROOF SHEATHING

FASCIA

PLYWOOD SOFFIT

HORIZONTAL SOFFIT

RAFTER

TOP PLATE

STUD

CONTINUOUS NAILER

OUTRIGGER

PLYWOOD ROOF SHEATHING

INTERMITTENT SCREENED SOFFIT VENTS

PLYWOOD ROOF SHEATHING

FASCIA

PLYWOOD SOFFIT

SCREEN VENT RUNS LENGTH OF SOFFIT

STANDARD HORIZONTAL SOFFIT—SLOPING ROOF

STANDARD SLOPING SOFFIT AND ROOF

FLAT ROOF WITH SLOPING SOFFIT—NOTE VENT STRIP

Reproducible Master 13-2 (Part A)
Siding Application

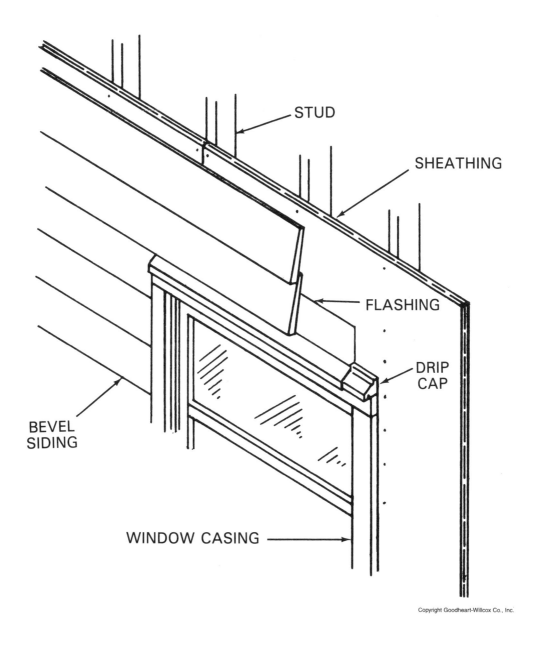

STUD

SHEATHING

FLASHING

DRIP
CAP

BEVEL
SIDING

WINDOW CASING

Reproducible Master 13-2 (Part B)
Siding Application

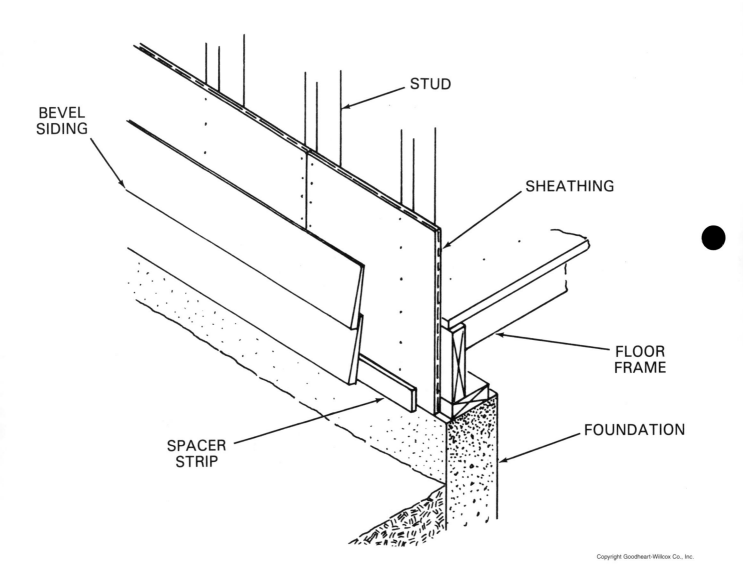

BEVEL
SIDING

STUD

SHEATHING

FLOOR
FRAME

SPACER
STRIP

FOUNDATION

Typical Exterior Insulation
Finish System

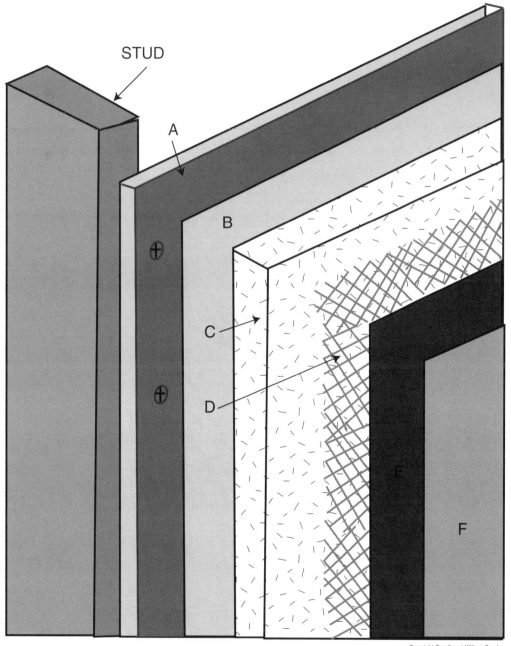

Components of an EIFS-sided structure. These materials may be applied over any of several different substrates: Plywood, cement board, oriented strand board, glass-faced, etc. A—Cement board. B—Acrylic copolymer adhesive or polymer modified cementitious adhesive. C—Expanded polystyrene insulation board. D—Open weave glass fiber reinforcing mesh. E—Exterior base coat of Portland cement mortar containing dry latex polymers. F—Surface coating based on an acrylic polymer emulsion. (U.S. Gypsum Co.)

Reproducible Master 13-4

Brick Veneer Construction

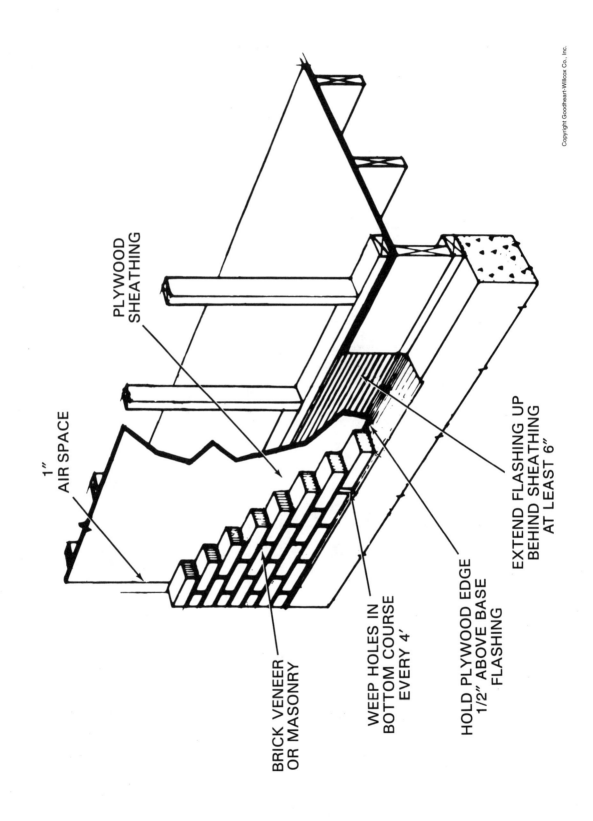

PLYWOOD SHEATHING

1" AIR SPACE

EXTEND FLASHING UP BEHIND SHEATHING AT LEAST 6"

BRICK VENEER OR MASONRY

WEEP HOLES IN BOTTOM COURSE EVERY 4'

HOLD PLYWOOD EDGE 1/2" ABOVE BASE FLASHING

Unit 13 Procedure Checklist
Installing Horizontal Wood Siding

Name _____ Total _____

☐ Properly prepares the wall surface with sheathing material; installs flashing where required.   5   4   3   2   1   Neglects to install sheathing material and flashing.

☐ Prepares story pole with appropriate measurements; transfers these measurements to corners; lays out measurements around doors and windows.   5   4   3   2   1   Tries to lay out measurements with tape measure, possibly resulting in inaccurate measurements.

☐ Installs spacer strip along the foundation wall; installs first course of siding, allowing lower edge to extend below spacer strip.   5   4   3   2   1   Neglects to install a spacer strip; installs first course with lower edge even with, or above the spacer strip.

☐ Determines method to be used for inside corners; carefully cuts and installs additional sections of siding.   5   4   3   2   1   Neglects to make adjustments for inside corners; raggedly cuts additional sections of siding.

☐ Completes installation of wood siding with coat of water-repellent preservative.   5   4   3   2   1   Fails to coat structure with preservative.

Unit 13 Quiz
Exterior Wall Finish

Name _____ Score _____

True-False

Circle T if the answer is True or F if the answer is False.

T   F   1. Most shingles are made in random widths.

T   F   2. Plywood siding is made from interior-grade plywood.

T   F   3. Hardboard siding panels are available in standard 4′ lengths.

T   F   4. Balloon framing should be utilized if a stucco finish is to be applied.

Multiple Choice

Choose the answer that correctly completes the statement. Write the corresponding letter in the space provided.

_____   5. The _____ is the main trim member along the edge of a roof.

          A. cornice
          B. rake
          C. fascia board
          D. None of the above.

_____   6. Eight or ten inch wood siding should overlap about _____ .

          A. 1 1/2″
          B. 2″
          C. 3 1/2″
          D. 5″

_____   7. The standard length for shingle panels is _____ .

          A. 4′
          B. 6′
          C. 8′
          D. 10′

_____   8. _____ bundles of shingles equal one square.

          A. Two
          B. Four
          C. Six
          D. Eight

Identification

Identify the parts of the EIFS-type siding material.

_____ 9. A

_____ 10. B

_____ 11. C

_____ 12. D

_____ 13. E

_____ 14. F

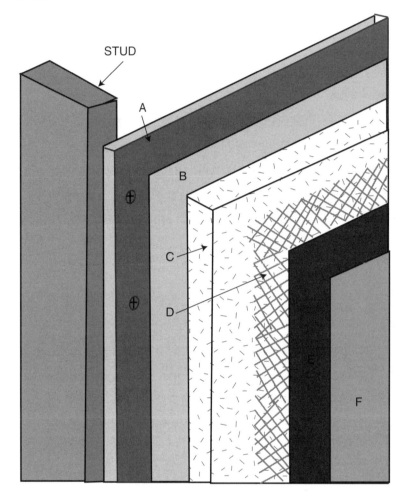

Section 3 Exam
Closing In

Name _____ Score _____

True-False

Circle T if the answer is True or F if the answer is False.

T   F   1.   For shingle installation, most manufacturers recommend 12 gauge galvanized nails with barbed shanks.

T   F   2.   One-piece flashing is required around chimneys.

T   F   3.   Any type of finish nail can be used when installing wood shakes.

T   F   4.   A material with a high R-value has a great amount of resistance to heat passage.

T   F   5.   When listing the dimensions of windows, always state the horizontal dimension first.

T   F   6.   Most window units are installed from the outside.

T   F   7.   Sheathing paper is applied directly to the wall frame to resist air and moisture infiltration when sheathing is not used.

T   F   8.   Ice and water barrier should extend 24″ above the exterior wall line.

T   F   9.   In residential construction, exterior doors usually swing inward.

T   F   10.   Less than 1/4″ clearance should be allowed for windows.

Multiple Choice

Choose the answer that correctly completes the statement. Write the corresponding letter in the space provided.

_____   11.   The trim board applied to the lower edge of a sloping roof before applying the surface material is called the _____.
   A. gutter
   B. drip edge
   C. fascia
   D. starter strip

_____   12.   The amount of roofing material needed to provide 100 square feet of coverage is called a _____.
   A. square
   B. bundle
   C. lot
   D. None of the above.

_____   13.   In frame construction, the horizontal location of windows is measured to the _____ of the window unit.
   A. left side
   B. centerline
   C. right side
   D. None of the above.

_____ 14. The purpose of a _____ is to divert the flow of water and prevent ice and snow build-up behind the chimney.

    A. saddle
    B. fascia
    C. drip edge
    D. starter strip

_____ 15. The term _____ refers to a waterway built into the roof surface over the cornice.

    A. gutter
    B. downspout
    C. fascia
    D. eaves trough

_____ 16. Main entrance doors are generally _____ wide.

    A. 2'-6"
    B. 3'-0"
    C. 3'-6"
    D. None of the above.

_____ 17. Underlayment _____.

    A. protects the sheathing from moisture until shingles are laid
    B. provides additional weather protection by preventing the wind-driven rain and snow from entering
    C. prevents contact between shingles and resinous areas in the sheathing
    D. All of the above.

_____ 18. Three-tab shingles require a minimum of _____ nails per strip.

    A. 2
    B. 3
    C. 4
    D. 6

_____ 19. _____ is often used for base flashing around chimneys.

    A. Asphalt-impregnated sheathing
    B. Sheet metal
    C. Cap flashing
    D. None of the above.

Matching

Select the correct answer from the list on the right and place the corresponding letter in the blank on the left.

_____ 20. Coverage.

_____ 21. Exposure.

_____ 22. Head lap.

_____ 23. Side lap.

_____ 24. Shingle butt.

    A. The lower, exposed edge of a shingle.
    B. The distance (in inches) between the edges of one course and the next higher course.
    C. Indicates the amount of weather protection provided by the overlapping of shingles.
    D. The distance (in inches) from the lower edge of an overlapping shingle to the top edge of the shingle beneath.
    E. The overlap length (in inches) for side-by-side roofing elements.

Name _____

_____ 25. Double-hung window.

_____ 26. Casement window.

_____ 27. Awning window.

_____ 28. Hopper window.

_____ 29. Jalousie window.

A. Sash is hinged on the side and swings outward.
B. Sash is hinged at the top and swings out at the bottom.
C. Consists of two sash that slide up and down in the window frame.
D. Consists of a series of horizontal glass slats held at each end by a movable metal frame.
E. Sash is hinged along the bottom and swings inward.

Completion

Place the answer that correctly completes the statement in the space provided.

_____ 30. _____ flashing is commonly used to waterproof joints between sloping roofs and vertical walls. Step

_____ 31. The minimum roof slope for wood shakes is _____ in 12.

_____ 32. Concrete roofing tile usually weigh from _____ to _____ lb. per sq. ft.

_____ 33. Common glass is produced by a process called _____.

_____ 34. In residential construction, the standard height of windows measured from the bottom side of the head to the finished floor is _____.

_____ 35. The acronym EIFS stands for the siding system _____ .

Identification

Identify the parts of exterior wall finish.

_____ 36. Bevel siding.

_____ 37. Drip cap.

_____ 38. Stud.

_____ 39. Window casing.

_____ 40. Sheathing.

_____ 41. Flashing.

Identification

Identify the types of horizontal siding.

_____ 42. Rabbeted bevel siding.

_____ 43. Drop siding.

_____ 44. Rustic siding.

_____ 45. Clapboards.

_____ 46. Bevel siding.

_____ 47. Log cabin.

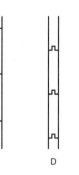

A   B   C   D   E   F

Thermal and Sound Insulation

OBJECTIVES

Students will be able to:

❑ Summarize the principles of conduction, convection, and radiation.

❑ Define technical terms relating to thermal and acoustical properties of construction materials.

❑ Interpret thermal ratings charts.

❑ Describe the types of insulation available today.

❑ Select appropriate areas for insulation in a given structure.

❑ Explain the principle of condensation.

❑ Describe methods of controlling moisture problems.

❑ List general procedures for installing batt and blanket, fill, and rigid insulation.

❑ Define acoustical terms.

❑ Describe methods of construction that raise STC ratings in desired areas.

INSTRUCTIONAL MATERIALS

Text: Pages 393-428

Important Terms, page 426

Test Your Knowledge, page 426

Outside Assignments, page 427

Workbook: Pages 97-106

Instructor's Manual:

Reproducible Master 14-1, *Typical Insulating Values*

Reproducible Master 14-2, *Insulating Band Joists and Ceiling, Parts A, B and C*

Procedure Checklist—*Installing Batts and Blankets*

Unit 14 Quiz

TRADE-RELATED MATH

The amounts of insulating materials are calculated on the basis of the area needed to fill. To determine the amount of insulating material needed for exterior walls, first calculate the perimeter of the structure. Then, multiply by the ceiling height. Subtract the area of any window or door openings. This net area is then divided by the number of square feet in a roll or batt. For example, if the perimeter of the house is 400′, the ceiling height 8′ and windows and doors comprise 96 sq. ft., determine the net wall area.

$$\text{Net wall area} = (\text{perimeter} \times \text{ceiling height}) - \text{area of openings}$$
$$= (400 \times 8) - 96$$
$$= 3200 - 96$$
$$\text{Net wall area} = 3104 \text{ sq. ft.}$$

If 3″ blanket insulation (70 square feet per roll) is used, how many rolls will be required?

$$\text{Number of rolls} = \text{net wall area} \div \text{sq. ft./roll}$$
$$= 3104 \div 70$$
$$\text{Number of rolls} = 45 \ (44\frac{1}{3} \text{ approximately})$$

INSTRUCTIONAL CONCEPTS AND STUDENT LEARNING EXPERIENCES

BUILDING SEQUENCE

1. Have the students list some of the other tradespeople that are involved in building a residential structure. Have them explain what types of activities the tradespeople perform. Stress that a carpenter should check to see if additional reinforcing of the framing members is needed after their activities have been completed on a job site.

HOW HEAT IS TRANSMITTED

1. Discuss the three methods used to transfer heat—conduction, convection, and radiation.

2. Describe the concept of conduction as it relates to building materials.

3. Describe the concept of convection, stressing that

this type of heat transfer only occurs in fluids or gases.

4. Describe the concept of radiation as it relates to building components.

THERMAL INSULATION

1. Have students list materials that they feel are good or poor insulators. Explain that the materials they listed as good insulators should be fairly porous materials, and materials they listed as poor insulators should be fairly dense. Discuss the molecular composition of these materials and how it affects insulating qualities.

2. Cite the types of materials that are commonly used as insulators including glass fibers, glass foam, mineral fibers, organic fibers, and foamed plastic.

3. Describe the characteristics of good insulating materials.

HEAT LOSS COEFFICIENTS

1. Define the following terms: Btu, k, C, R, and U as they relate to building materials.

2. Using Reproducible Master 14-1, *Typical Insulating Values,* discuss the R-values and U-values of the materials that are shown.

3. Refer to Figure 14-6 and have students note the improved insulating values as they progress down the chart.

4. Referring to Figure 14-7, discuss insulating qualities of different types of materials and the implications for construction.

HOW MUCH INSULATION

1. List factors that should be considered when determining the amount of insulation required for a structure.

2. Refer to Figure 14-8 and have students determine the amount of insulation required for ceilings, walls, and floors in your specific location. Emphasize that insulation not only provides protection against the cold, but also helps to keep the heat out in the summer.

3. Define "degree day," and discuss how it is calculated. Have the students find out the number of degree days per year in your area of the country.

4. Review the relationship between U-values and R-values. Emphasize that heat transmission decreases as the thickness of insulation increases, but not in a direct relationship. It is important for

the students to understand that at some point, additional insulation will not make a significant difference in the amount of heat transmission.

TYPES OF INSULATION

1. Identify the five major classifications of insulation—flexible, loose fill, rigid, foamed, and reflective. Obtain samples of these types of insulation for student inspection.

2. Describe the composition of blanket and batt insulation.

3. Distinguish between blanket and batt insulation.

4. Discuss the composition of loose fill insulation, and describe its purpose in building construction.

5. Describe the composition of rigid insulation. List places where rigid insulation is commonly installed in houses.

6. Discuss advantages/disadvantages of blown-in and foamed-in-place insulation materials.

7. Discuss the use of reflective insulation in residential construction. Emphasize that it is only effective when the reflective material is exposed to an air gap of 3/4" or more.

OTHER TYPES OF INSULATION

1. Discuss other types of insulation that are used in residential construction. If possible, have some of these types of insulation on hand for student inspection.

WHERE TO INSULATE

1. Identify areas of a house that should be insulated. Emphasize that the insulation should be placed as close to the heated area as possible.

2. Have students explain the advantages of insulating a basement area.

3. Using part *A* of Reproducible Master 14-2, *Insulating Band Joists & Ceiling,* discuss the advantages of insulating along a band joist.

4. Describe how crawl spaces should be insulated, emphasizing the placement of the vapor barrier.

5. Explain how the ground under a crawl space should be covered to control the amount of moisture entering the building.

6. Discuss the insulation required for a slab-on-grade structure. Mention that only the perimeter need be insulated since very little heat is lost through the ground under the center of the structure.

7. Using Figures 14-23 and 14-24, describe how insulation is added to slabs-on-grade and existing

foundation walls. Point out the use of flashing in adding insulation to the outside of existing foundation walls.

CONDENSATION

1. Discuss the problems that may occur because of condensation in a residential structure.
2. Define the term "dew point," and explain how it relates to the construction of a house.

VAPOR BARRIERS

1. Emphasize the proper placement of a vapor barrier to avoid condensation in walls.
2. Identify types of materials that are used as vapor barriers. Mention that many modern insulating materials already have a vapor barrier applied to one of the surfaces.
3. Discuss when vapor barriers should be applied. Stress that any punctures or cuts in a vapor barrier will reduce its effectiveness.

VENTILATION

1. Explain the importance of having good ventilation in a structure. Cite possible scenarios that may occur if less-than-adequate ventilation is used.
2. Emphasize the importance of not blocking the airway near the soffit in an attic when installing thicker insulation. Describe possible means to ensure that the airway is not blocked.

SAFETY WITH INSULATION

1. Emphasize the importance of wearing appropriate clothing—tight-fitting cap, gloves, long-sleeve shirt, and long pants—when installing any type of insulation. In addition, a mask covering the mouth and nose should also be worn to avoid inhaling resin fibers.
2. Stress the need for good housekeeping in reference to proper disposal of fiberglass scraps.

INSTALLING BATTS AND BLANKETS

1. Describe how blankets or batts can be laid out and cut on the floor.
2. Explain how batts are installed in wall sections. Mention that if batts must be joined, there should be at least a 1″ overlap between the pieces. Describe the three ways that flanges might be stapled to hold the insulation into position. If possible, demonstrate these methods.
3. Describe how insulation is installed in a situation

where drywall will be placed over it.
4. Discuss how insulation is installed around electrical boxes and plumbing pipes. Demonstrate the correct procedures.
5. Using part *B* of Reproducible Master 14-2, *Insulating Band Joists and Ceiling*, describe how insulation is installed in a ceiling. Have students note that the insulation is held back from the end of the rafter to allow sufficient room for airflow.
6. Using part *C* of Reproducible Master 14-2, *Insulating Band Joists and Ceiling,* explain how insulation is installed along the floor frame of a multi-story structure.
7. Discuss the importance of insulating around window and door frames. Mention that cuttings left over from other pieces can be stuffed into these places and then covered with a vapor barrier.

INSTALLING FILL INSULATION

1. Discuss the two methods used to install fill insulation.
2. Describe the procedure used to pour fill insulation. Emphasize that a vapor barrier should be installed prior to installing the insulation. Demonstrate this procedure if time and space allows.
3. Explain the purpose of installing batt insulation around the perimeter of a ceiling prior to installing fill insulation.
4. Discuss the procedure used to blow fill insulation. Stress the importance of using protective equipment such as a face mask, goggles, and gloves.

INSTALLING RIGID INSULATION

1. Discuss applications for rigid insulations.
2. Explain the purpose of applying a plaster coat after installing rigid insulation to an exterior masonry wall.

INSULATING BASEMENT WALLS

1. Describe the preparation necessary before installing batt insulation on basement walls. Once again, emphasize the importance of installing a vapor barrier on the warm side of the insulation.

INSULATING EXISTING STRUCTURES

1. Discuss the concerns that should be addressed when insulating an existing structure. Describe materials that can be used instead of a conventional vapor barrier.

AIR INFILTRATION

1. Discuss the locations of a structure that should be caulked or sealed. List the materials that can be used to caulk these locations. Emphasize the importance of sealing such places as electrical conduit and plumbing runs that extend from the attic or basement.

ESTIMATING MATERIALS

1. Describe how to estimate the amount of insulation for exterior walls.
2. Describe how to estimate the amount of insulation for floors and ceilings.

ACOUSTICS AND SOUND CONTROL

1. Have the students identify the possible sources of noise in a residential structure.

ACOUSTICAL TERMS

1. Define the following acoustical terms as they relate to sound control in a residential structure: sound, decibel, reverberation sounds, frequency, impact sounds, masking sounds, decibels reduction, sound transmission loss (TL), sound absorption, noise reduction coefficient (NRC), and sound transmission class (STC).
2. Using Figure 14-57, have students identify the decibel levels of sounds and noises that are familiar to them.
3. Describe the action of sound when it is generated within a room. Discuss the diaphragm action that occurs, allowing sound to be transmitted through a wall.
4. Define the term "masking sounds" as it relates to designing a sound insulating panel.

WALL CONSTRUCTION

1. Referring to Figure 14-61, explain the methods of constructing interior walls and the sound transmission classes assigned to them.
2. Describe the composition of sound deadening board and how it can be used to increase the STC of a partition. Explain how sound deadening board is identified.
3. Discuss the use of double walls to increase the STC. Describe the construction of double walls using resilient channels.
4. Compare the methods used for soundproofing walls to floors and ceilings.
5. Describe the use of spring clips to attach the ceiling material.

6. Describe a method used to soundproof an existing floor using sleepers and glass wool. Emphasize that the sleepers "float" on the glass wool and are not actually tied into the existing floor.
7. Identify materials that can be used to reduce the amount of sound transmitted around a door.
8. Have students identify the rooms of a house in which sound control is most important. Describe materials that can be used to reduce the amount of sound transmission.
9. Discuss how acoustical materials, such as acoustical tiles, dissipate sound energy.
10. Briefly describe how acoustical materials are installed.
11. Discuss how suspended ceilings are used to reduce the amount of sound transmission.

INSTALLATION OF MATERIALS

1. Emphasize that the manufacturer's installation recommendations should be followed to ensure that acoustical materials work properly.
2. Describe the preparation necessary before painting perforated boards or acoustical tile. Mention that these materials can be painted as long as the pores are not clogged with paint.

UNIT REVIEW

1. Review the unit objectives. Be sure that the students fully understand each objective.
2. Assign *Important Terms, Test Your Knowledge* questions, and *Outside Assignments* on pages 426-427 of the text. Review the answers in class.
3. Assign pages 97-106 of the workbook. Review the answers in class.

EVALUATION

1. Use Procedure Checklist—*Installing Batts and Blankets,* to evaluate the techniques that students used to perform these tasks.
2. Use Unit 14 Quiz for in-class evaluation. Correct the quizzes and return them to the students for review.

ANSWERS TO TEST YOUR KNOWLEDGE TEXT PAGE 426

1. conduction
2. radiation
3. durability
4. R
5. See page 396, Figure 14-6.
6. 24″, 48″

7. dew point
8. R-49
9. warm
10. top
11. blown
12. Rigid
13. Expanded polyurethane.
14. Refer to page 402, Figure 14-16.
15. 8-10
16. infiltration
17. decibel
18. 30
19. Sound Transmission Class or STC
20. cane
21. stud space
22. 3
23. Noise Reduction Coefficient

ANSWERS TO WORKBOOK QUESTIONS PAGES 97-106

1. heating
2. resistance
3. conduction
4. A. Radiation.
   B. Convection.
5. moisture proof
6. A. hour
   B. 1″
7. 1056 Btu.
8. B. U = 0.05
9. C. R-14
10. 0.087
11. R-20.8 to R-22.6
12. 65°F
13. reflective
14. A. Blanket.
    B. Batt.
    C. Loose fill.
15. A. foamed plastic
16. B. 3/4″
17. D. 55 lb.
18. A. Siding.

B. Flashing.
C. Extruded polystyrene.
19. dew point
20. C. between the inside wall covering and wall frame
21. C. Both the insulation and vapor barrier should be continuous under the entire floor.
22. D. 1/1600
23. B. 12″
24. C. polystyrene
25. A. pouring
    B. blowing
26. Walls: 840
    Ceiling: 1100
27. decibel
28. second
29. C. Noise Reduction Coefficient
30. 39 dB
31. D. Sound Transmission Class
32. A. 33
    B. 39
    C. 45
    D. 50
33. stud space
34. A. 20-25 dB
35. C. 70%
36. A. spray gun
37. A
38. A

ANSWERS TO UNIT 14 QUIZ

1. True.
2. True.
3. True.
4. False.
5. True.
6. D. All of the above.
7. B. Btu
8. D. Either B or C.
9. A. Impact
10. D. Either A or B.
11. C. dew point

Modern Carpentry Instructor's Manual

Reproducible Master 14-1
Typical Insulating Values

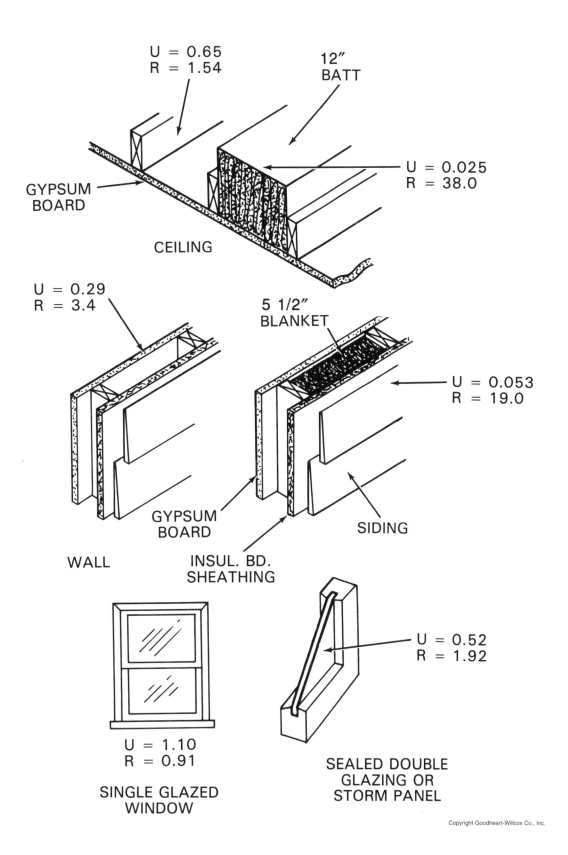

U = 0.65
R = 1.54

12"
BATT

U = 0.025
R = 38.0

GYPSUM
BOARD

CEILING

U = 0.29
R = 3.4

5 1/2"
BLANKET

U = 0.053
R = 19.0

GYPSUM
BOARD

SIDING

WALL

INSUL. BD.
SHEATHING

U = 0.52
R = 1.92

U = 1.10
R = 0.91

SINGLE GLAZED
WINDOW

SEALED DOUBLE
GLAZING OR
STORM PANEL

Reproducible Master 14-2 (Part A)
Insulating Band Joists & Ceilings

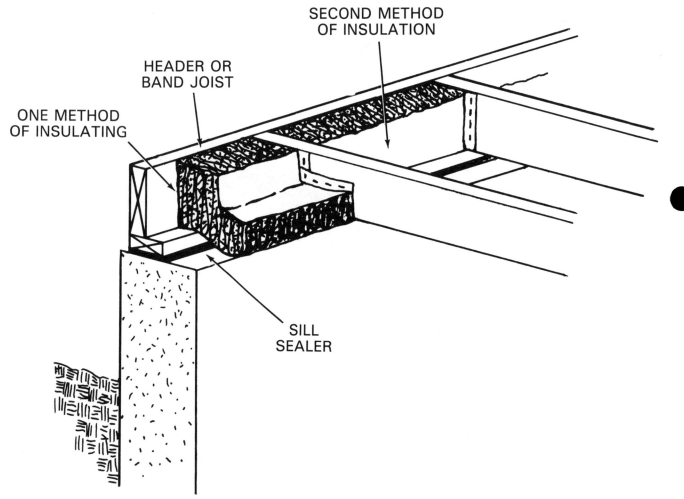

SECOND METHOD
OF INSULATION

HEADER OR
BAND JOIST

ONE METHOD
OF INSULATING

SILL
SEALER

Reproducible Master 14-2 (Part B)
Insulating Band Joists & Ceilings

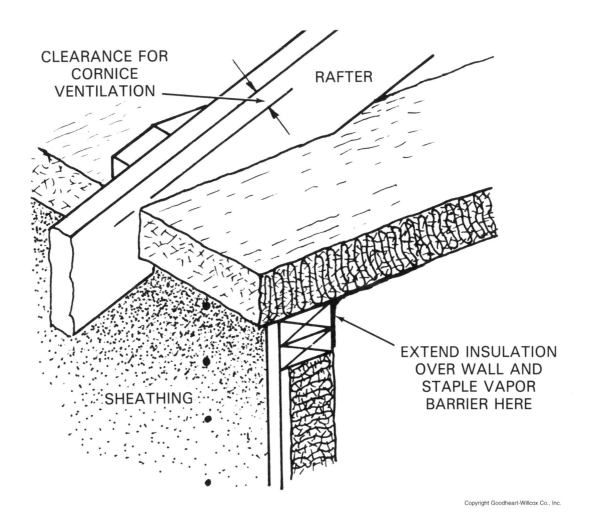

CLEARANCE FOR CORNICE VENTILATION

RAFTER

EXTEND INSULATION OVER WALL AND STAPLE VAPOR BARRIER HERE

SHEATHING

Reproducible Master 14-2 (Part C)
Insulating Band Joists & Ceilings

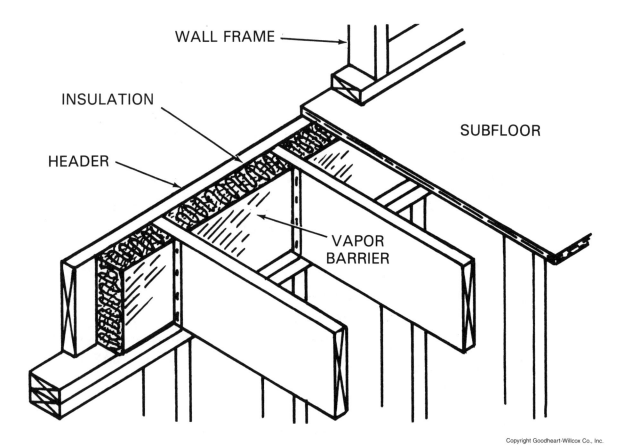

WALL FRAME

INSULATION

SUBFLOOR

HEADER

VAPOR
BARRIER

Unit 14 Procedure Checklist
Installing Batts and Blankets

Name _____   Total _____

❏ Wears proper protective equipment.	5	4	3	2	1	Neglects to wear protective equipment.
❏ Carefully measures and cuts batts or blankets.	5	4	3	2	1	Incorrectly measures batt or blanket length; carelessly cuts insulation resulting in wasted insulation.
❏ Properly installs batts by starting from the bottom and working up and at the top and working down; overlaps vapor barrier.	5	4	3	2	1	Incorrectly installs insulation; vapor barrier facing toward the exterior of the structure.
❏ Secures flanges in place (if necessary); applies separate vapor barrier (if required).	5	4	3	2	1	Neglects to secure flanges into place; forgets to apply separate vapor barrier (if required).

227

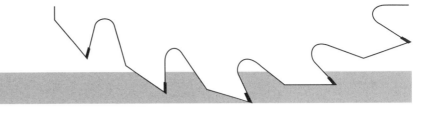

Unit 14 Quiz

Thermal and Sound Insulation

Name _____ Score _____

True-False

Circle T if the answer is True or F if the answer is False.

T F 1. Radiation is the process of heat transfer by means of wave motion.

T F 2. More insulation is required in a climate with a large number of degree days.

T F 3. A vapor barrier must be located on the warm side of an insulated wall.

T F 4. Frequency is the unit of measure used to indicate the loudness of a sound.

T F 5. A good insulating material has a high R-value.

Multiple Choice

Choose the answer that correctly completes the statement. Write the corresponding letter in the space provided.

_____ 6. Heat transfer takes place by means of _____.

    A. convection
    B. conduction
    C. radiation
    D. All of the above.

_____ 7. A(n) _____ is the amount of heat needed to raise the temperature of 1 lb. of water 1° Fahrenheit.

    A. calorie
    B. Btu
    C. R-value
    D. degree day

_____ 8. Moisture coming up through the ground can be controlled by covering the area with _____.

    A. loose fill insulation
    B. heavy roll roofing
    C. 6 mil polyethylene plastic film
    D. Either B or C.

_____ 9. _____ sounds are carried throughout a building by the vibrations of the structural materials.

    A. Impact
    B. Reverberation
    C. Masking
    D. None of the above.

_____ 10. Acoustical plaster should be _____ onto the surface.

    A. troweled
    B. spread
    C. sprayed
    D. Either A or B.

_____ 11. The _____ is the temperature at which the air is completely saturated with moisture.
   A. R-value
   B. U-value
   C. dew point
   D. saturation point

15

Interior Wall and Ceiling Finish

Students will be able to:

- ❏ Discuss wall and ceiling covering materials.
- ❏ Describe wallboard cutting, nailing, and adhesive techniques.
- ❏ Describe the characteristics of gypsum plaster.
- ❏ Explain how gypsum and metal lath are applied.
- ❏ Illustrate the use of plaster grounds.
- ❏ Describe plastering methods.
- ❏ Illustrate how double layer and predecorated wallboard are applied.
- ❏ List the procedures for installing wood paneling.
- ❏ Lay out ceiling tile and install furring strips.
- ❏ Describe methods for leveling and installing a suspended ceiling.
- ❏ Estimate quantities of lath, wallboard, and ceiling tiles for a specific interior.

INSTRUCTIONAL MATERIALS

Text: Pages 429-462

    Important Terms, page 461

    Test Your Knowledge, page 461

    Outside Assignments, page 462

Workbook: Pages 107-116

Instructor's Manual:

    Reproducible Master 15-1, *Drywall Application, Parts A* and *B*

    Reproducible Master 15-2, *Suspended Ceiling System*

    Procedure Checklist—*Installing Drywall*

    Unit 15 Quiz

TRADE-RELATED MATH

    Plasterers usually base their prices and estimates on the number of square yards to be covered. Square yards can be determined by dividing the number of square feet by 9. For example, to determine the number of square yards contained in a wall measuring $8' \times 24'$, the following calculations would be needed:

$$\begin{aligned} \text{Square feet} &= \text{height} \times \text{width} \\ &= 8' \times 24' \\ &= 112 \text{ sq. ft.} \end{aligned}$$

$$\begin{aligned} \text{Square yards} &= \text{square feet} \div 9 \\ &= 112 \div 9 \\ &= 12.4 \text{ sq. yd.} \end{aligned}$$

INSTRUCTIONAL CONCEPTS AND STUDENT LEARNING EXPERIENCES

DRYWALL INSTALLATION

1. Have the students list the types of materials used to cover interior walls. Have as many of these materials on hand as possible for student inspection. Discuss the common sizes of each type of material.

2. Discuss the advantages and disadvantages of drywall or plaster as interior wall finish materials. Mention that drywall is the most prominent type of interior wall finish material used today.

3. Identify types of gypsum wallboard that are not suitable as interior wall finish materials.

4. Describe the difference between single layer and double layer drywall construction. Cite advantages and disadvantages of each type of construction.

5. Using part *A* of Reproducible Master 15-1, *Drywall Application*, describe the two methods of installing drywall sheets. Note advantages of each method.

6. Discuss the techniques used to lay out and make straight drywall cuts. Stress the importance of using a sharp knife.

7. Demonstrate the proper techniques for cutting drywall. Show students how to make straight cuts as well as curved cuts.

8. Identify the various types of nails and screws that can be used to attach drywall to studs or other framing.

9. Discuss the nailing pattern and spacing that

should be used for drywall. Emphasize the importance of drawing the drywall tight against the framing.

10. Demonstrate the correct methods used to install sheets of drywall on walls. Show the students how to drive nails so as to leave a dimple on the surface of the drywall without breaking the paper covering. In addition, show the correct methods for installing drywall with screws.

12. Demonstrate the correct methods of installing drywall on a ceiling. Discuss possible problems that may be encountered when installing drywall overhead.

13. Discuss the procedure for installing drywall with an adhesive. Cite applications of such an installation.

14. Discuss how joint compound and reinforcing tape are applied over seams and fasteners. Demonstrate the proper procedure in detail.

15. Describe the use of pressure-sensitive glass-fiber tape to cover the seams of drywall.

16. Discuss the use of a taping tool to apply reinforcing tape. Cite advantages over the other method of applying tape.

17. Identify the materials used to reinforce inside and outside corners of drywall. Demonstrate the proper procedure for installing corner beads and paper flanges.

DOUBLE LAYER CONSTRUCTION

1. Using part *B* of Reproducible Master 14-1, *Drywall Application*, discuss double layer application. Cite advantages of this method over the single layer application.

2. Discuss the means for attaching the base layer to the framing and finish layer to the base layer. Emphasize that the joints of the finish layer should be offset from those of the base layer by at least 10″.

FINISHING DOUBLE LAYER WALLBOARD

1. Compare the finishing of double layer wallboard to that of single layer construction.

MOISTURE RESISTANT (MR) WALLBOARD AND CEMENT BACKERBOARD

1. Distinguish between common drywall, MR wallboard, and cement backerboard. Explain where MR wallboard and cement backerboard should be used.

2. Referring to Figure 15-32, discuss the installation

of MR wallboard or cement board around the perimeter of a bathtub. Stress the importance of avoiding application of regular joint compound on any surfaces to be tiled.

3. Referring to Figures 15-29 through 15-31, discuss tools and methods used when working with cement board.

4. Mention that the correct placement of holes for the valve handle(s) and spout can be ensured by first laying out the dimensions on a large piece of cardboard, and then fitting the cardboard into position. The cardboard can be removed and the layout transferred to the MR wallboard.

VENEER PLASTER

1. Describe the application of veneer plaster. Emphasize that the manufacturer's directions should be followed carefully for the specific type of plaster being used.

PREDECORATED WALLBOARD

1. Discuss the use of predecorated wallboard. Mention that the wallboard is usually installed vertically because of the difficulty in matching and finishing butt joints.

2. Stress the importance of using a plastic-headed hammer, rawhide mallet, or a special cover placed over the face of a regular hammer when installing colored nails that match the surface.

3. Explain how edges and joints are covered when using predecorated wallboard.

WALLBOARD ON MASONRY WALLS

1. Describe the installation of drywall over masonry walls. Mention that the drywall can be attached directly to the masonry if the wall is straight and true, otherwise furring strips must first be applied, and then the drywall attached to it.

2. Explain how insulation should be installed on exterior masonry walls prior to installing drywall.

INSTALLING PLYWOOD

1. Discuss the preparation and storage methods that should be used for interior plywood.

2. Stress the importance of carefully planning the panel arrangement. Discuss the application designs shown in Figure 15-39.

3. Review portable saw safety. Mention that the cutting should be performed from the back side of the panels to avoid splintering.

4. Describe the installation of plywood panels over a masonry wall.

5. Explain the purpose of installing 1/4″ plywood over a base of 1/2″ drywall.

HARDBOARD

1. Discuss the special preparation that should be used for hardboard panels. Cite problems that may occur if the panels have not been allowed to acclimate to the humidity level.

PLASTIC LAMINATES

1. Explain why plastic laminates are generally bonded to backer material before installation on a wall.

SOLID LUMBER PANELING

1. Identify the types of softwood lumber that are commonly used as wall paneling. List the dimensions in which they are usually available.
2. Explain that furring strips are usually not required when wood paneling is applied horizontally.
3. Discuss how narrow widths of tongue and groove paneling are blind nailed to conceal the nail heads. Stress that the paneling should be allowed to attain a humidity level close to the humidity of the room in which it will be installed *before* installation.

PLASTER

1. List the wall and ceiling qualities that can be obtained from using plaster.
2. Identify the types of plaster bases used in modern residential construction. State the thicknesses that should be used for different stud spacing.
3. Describe the installation of plaster bases. Emphasize that the end joints should not be aligned when installing plaster bases.

METAL LATH

1. Cite the applications of metal lath in residential construction.
2. Describe the installation of metal lath and water-proof felt paper in areas subjected to moisture.

REINFORCING

1. Identify places where expanded metal lath is used to minimize cracking. Stress that the lath should be lightly tacked into position so as to become a part of the plaster base only.
2. Identify the type of reinforcement used for outside corners. Describe the purpose of corner beads.

PLASTER GROUNDS

1. Discuss the purpose of plaster grounds. Identify the types of materials commonly used as plaster grounds.

PLASTER BASE ON MASONRY WALLS

1. List the methods that can be used to attach furring strips to a masonry surface.
2. Discuss why furring strips would be used on a masonry surface when applying plaster. Describe the installation of insulation before applying plaster.

PLASTERING MATERIALS AND METHODS

1. Discuss the composition of plaster.
2. Distinguish between "two-coat work" and "three-coat work."
3. Describe the procedure for applying the scratch coat. Emphasize the thickness of the coat that should be obtained.
4. Describe the application of the brown coat. Emphasize the thickness of the coat that should be obtained.
5. Describe the application of the finish coat in three-coat work. Stress that the finish coat should only be about 1/16″ thick.
6. Demonstrate the application of plaster using the three-coat method.

CEILING TILES

1. List the types of ceiling tile that are commonly used in residential work.
2. Discuss the layout procedure used for ceiling tile. Emphasize that the border courses along the opposite walls should be the same width.
3. Discuss the purpose of using furring strips for ceiling tiles. Describe where the strips should be placed. Explain how the furring strips can be leveled.
4. Describe "sheet furring" and how it can be used in place of furring strips.
5. Emphasize the importance of checking the placement and levelness of furring strips prior to installing tile.
6. Describe the installation procedure for tile. Stress that the students should keep their hands clean while installing the tile to avoid getting dirt or smudges on them. Demonstrate the installation procedure for ceiling tile.
7. Discuss how a metal track system can be used for the installation of ceiling tiles.

SUSPENDED CEILINGS

1. Using Reproducible Master 15-2, *Suspended Ceiling System,* identify the parts of a suspended ceiling system. List advantages of using this type of system.
2. Describe the installation procedure for suspended ceilings. If possible, demonstrate the procedure.

ESTIMATING MATERIALS

1. Describe the method for estimating the amount of wall or ceiling materials.
2. Describe the method for estimating the amount of gypsum lath used on a residential project.

UNIT REVIEW

1. Review the unit objectives. Be sure that the students fully understand each objective.
2. Assign *Important Terms, Test Your Knowledge* questions, and *Outside Assignments* on pages 461-462 of the text. Review the answers in class.
3. Assign pages 107-116 of the workbook. Review the answers in class.

EVALUATION

1. Use Procedure Checklist—*Installing Drywall,* to evaluate the techniques that students used to perform these tasks.
2. Use Unit 15 Quiz for in-class evaluation. Correct the quizzes and return them to the students for review.

ANSWERS TO TEST YOUR KNOWLEDGE
TEXT PAGE 461

1. See page 429. Gypsum wallboard or drywall, wallboard for veneer plastering, predecorated gypsum paneling, plywood and particle board, hardboard and fiberboard, solid wood paneling, plaster, clay finishes, plastic laminates (any nine).
2. 1/2″, 5/8″
3. moisture resistant
4. ceilings
5. 2
6. 10
7. 1/4″, 7/16″, 1/2″
8. 8% to 10%
9. 48
10. fire resistant
11. 48″
12. metal lath
13. grounds
14. sand
15. second
16. 1/2
17. True.
18. Diagonal, chevron, and herringbone.
19. 2
20. 9
21. wires
22. square yards

ANSWERS TO WORKBOOK QUESTIONS
PAGES 107-116

1. A. 4′
   B. 14′
2. A. 12″
   B. 16″
3. A. 7″
   B. 8″
4. Cement board
5. D. T&G
6. A. Bedding coat.
   B. Reinforcing tape.
   C. 1/2″-3/4″
   D. 2″
7. B. 10″
8. C. green
9. B. 24
10. A. 1/16″
11. C. Moldings and panels are applied to the surface at the same time.
12. 48 hours
13. B. 8% to 10%
14. A. Eliminates need for vapor barrier.
15. A. 1/32″
16. D. 48″
17. D. 16 × 48
18. C. 1/2″
19. B. Use a No. 13 ga. nail with a minimum length of 1″.
20. grounds
21. A. Metal lath.
    B. Cornerite.
    C. Corner bead.
22. furring strips
23. A. Does not withstand moisture well.
24. A. scratch coat
    B. brown coat
25. A. In most residential plastering, the first two coats are applied almost simultaneously.
26. True.
27. B. unit weight
28. A. Stapling flange.

B. Tongue.

C. Groove.

29. C. cut notches in lower edge of low joists
30. 232 bd. ft.
31. A. in any one of the corners
32. metal track
33. Panels: 6

192 sq. ft.
34. sq. ft.: 4036

Bundles: 64
35. sq. yd.: 449
36. Tile: 300

Cartons: 6

ANSWERS TO UNIT 15 QUIZ

1. False.
2. False.
3. True.
4. False.
5. True.
6. True.
7. True.
8. B. knife
9. C. MR
10. D. All of the above.
11. C. fiberboard
12. D. Either A or B.

Reproducible Master 15-1 (Part A)
Drywall Application

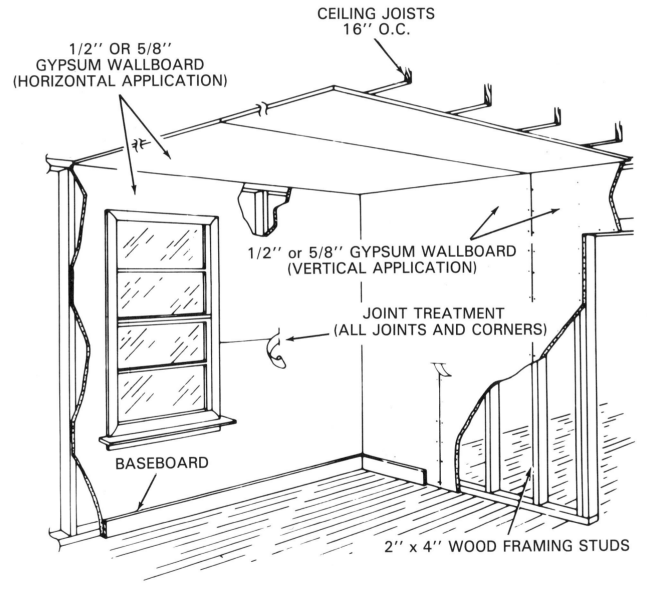

CEILING JOISTS
16'' O.C.

1/2'' OR 5/8''
GYPSUM WALLBOARD
(HORIZONTAL APPLICATION)

1/2'' or 5/8'' GYPSUM WALLBOARD
(VERTICAL APPLICATION)

JOINT TREATMENT
(ALL JOINTS AND CORNERS)

BASEBOARD

2'' x 4'' WOOD FRAMING STUDS

Reproducible Master 15-1 (Part B)
Drywall Application

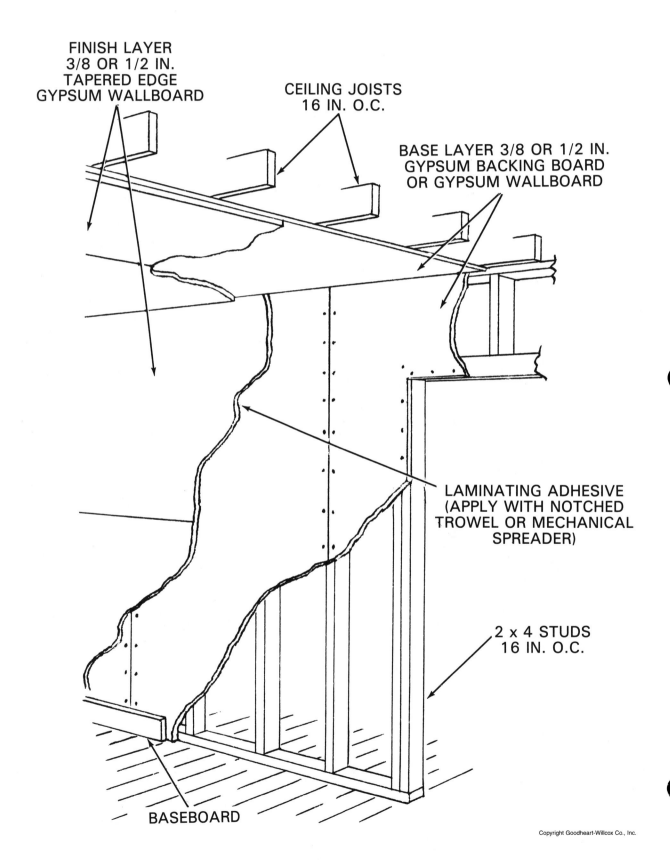

FINISH LAYER
3/8 OR 1/2 IN.
TAPERED EDGE
GYPSUM WALLBOARD

CEILING JOISTS
16 IN. O.C.

BASE LAYER 3/8 OR 1/2 IN.
GYPSUM BACKING BOARD
OR GYPSUM WALLBOARD

LAMINATING ADHESIVE
(APPLY WITH NOTCHED
TROWEL OR MECHANICAL
SPREADER)

2 x 4 STUDS
16 IN. O.C.

BASEBOARD

Reproducible Master 15-2

Suspended Ceiling System

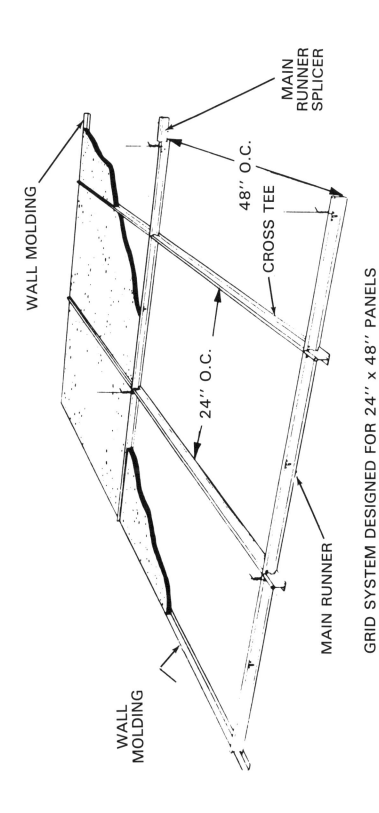

WALL MOLDING

MAIN RUNNER SPLICER

48" O.C.

CROSS TEE

24" O.C.

WALL MOLDING

MAIN RUNNER

GRID SYSTEM DESIGNED FOR 24" x 48" PANELS

Unit 15 Procedure Checklist
Installing Drywall Using Single Layer Construction

●

Name _____ Total _____

	5	4	3	2	1	
❑ Determines whether long edges will be parallel to or perpendicular with the studs; carefully makes measurements, taking a reading for each side of the panel;.	5	4	3	2	1	Disregards application direction for panels; takes only one measurement when laying out each panel.
❑ Accurately cuts panels using a knife; snaps panel by pressing down on the over hang; smoothes cut if necessary.	5	4	3	2	1	Uses hand saw or power saw to cut panel; produces uneven, inaccurate cut.
❑ Determines proper type of fastener; draws panels tight against studs; nails panel straight and true; "dimples" the nail head without breaking the surface.	5	4	3	2	1	Uses common nails for drywall; loosely fits panel against studs; breaks the surface of the panel when nailing.
❑ Applies bedding coat of joint compound to joints; embeds reinforcing tape into joint; applies skim coat over bedding boat; applies joint compound to fastener heads.	5	4	3	2	1	Neglects to apply bedding coat and reinforcing tape; forgets to cover fasteners.
❑ Applies another coat of joint compound after skim coat is completely dry, feathering the edges; applies final coat if necessary; sands all joints and fasteners.	5	4	3	2	1	Neglects to apply final coat of joint compound; does not feather the edges; forgets to sand joints and fasteners.

●

●

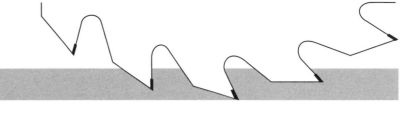

Unit 15 Quiz
Interior Wall and Ceiling Finish

Name _____ Score _____

True-False

Circle T if the answer is True or F if the answer is False.

T F 1. Particleboard is commonly referred to as drywall.

T F 2. When nailing drywall to studs, use a nail set to drive the heads of the nails below the surface.

T F 3. Drywall can be glued and/or nailed into position.

T F 4. It is not necessary to allow plywood to adjust to room temperature before using it since it is a "manufactured" product.

T F 5. When cutting plywood, cut from the back side of the panel.

T F 6. Cement board or backerboard is used only as floor underlayment.

T F 7. In three coat plaster work, the first coat is the scratch coat, the second coat is the brown coat, and the final coat is the finish coat.

Multiple Choice

Choose the answer that correctly completes the statement. Write the corresponding letter in the space provided.

_____ 8. Straight cuts should be made in drywall with a _____.

    A. reciprocating saw
    B. knife
    C. circular saw
    D. crosscut hand saw

_____ 9. _____ wallboard should be used as a base under ceramic tile in a shower.

    A. Backing
    B. Predecorated
    C. MR
    D. All of the above.

_____ 10. Why is it recommended to install 1/4″ plywood over a 1/2″ drywall base?

    A. To bring studs into alignment.
    B. To provide a rigid finished surface.
    C. To improve the fire-resistant qualities.
    D. All of the above.

_____ 11. Hardboard is also referred to as _____.

    A. drywall
    B. gypsum wallboard
    C. fiberboard
    D. particle board

_____ 12. In most modern construction, _____ is used as a base for plaster.
   A. gypsum lath
   B. expanded metal lath
   C. wood lath
   D. Either A or B.

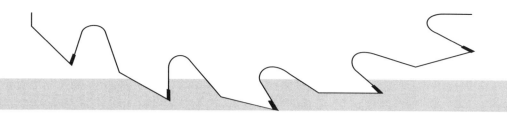

16

Finish Flooring

OBJECTIVES

Students will be able to:

❑ Describe strip, plank, and unit block wood flooring.

❑ Lay out and install strip flooring on concrete or plywood subfloors.

❑ Describe the procedure for applying hardboard, particleboard, waferboard, and plywood underlayment.

❑ Outline the basic steps for installing resilient flooring.

INSTRUCTIONAL MATERIALS

Text: Pages 463-482

Important Terms, page 480

Test Your Knowledge, page 481

Outside Assignments, page 481

Workbook: Pages 117-122

Instructor's Manual:

Reproducible Master 16-1, *Installing Strip Flooring, Parts A* and *B*

Reproducible Master 16-2, *Strip Flooring Sequence*

Reproducible Master 16-3, *Laying Out Parquet Flooring*

Procedure Checklist—*Installing Wood Strip Flooring*

Procedure Checklist—*Installing Resilient Flooring*

Unit 16 Quiz

TRADE-RELATED MATH

To determine the board feet of strip flooring needed to cover a given area, first calculate the area in square feet. Then add a percentage (shown in Figure 16-10) for the particular size being used. For example, if the floor dimensions are 20′ × 20′, and 3/4″ × 2″

wood strip flooring is used, the following calculations would be used:

$$\text{Total area} = \text{width} \times \text{length}$$
$$= 20 \times 20$$
$$\text{Total area} = 400 \text{ sq. ft.}$$
$$\text{Bd. ft. of flooring} = \text{floor area} + (\text{floor area} \times \text{percentage})$$
$$= 400 + (400 \times .425)$$
$$= 400 + 170$$
$$\text{Bd. ft. of flooring} = 570$$

INSTRUCTIONAL CONCEPTS AND STUDENT LEARNING EXPERIENCES

WOOD FLOORING

1. List the types of hardwoods and softwoods used for wood flooring. Explain why these woods are selected as flooring materials.

2. Have the students identify the three general types of wood flooring used in residential construction including strip, plank, and block.

3. Identify the common sizes and standard thicknesses of hardwood strip flooring. Mention that random widths are available for hardwood strip flooring.

4. List the range of widths available for hardwood plank flooring.

5. Discuss the factors that are considered when grading wood flooring including appearance, knots, streaks, color, pinworm holes, and sapwood. Have the students refer to page 751 of Appendix B, for additional information regarding the grading of wood flooring.

6. Discuss the purpose of having wood flooring delivered a few days before actual installation.

7. Describe the subflooring that is required for wood flooring.

8. Describe the features of strip flooring such as the

tongue and groove sides and ends, and the undercut. Discuss the purpose of these features.

9. Using part *A* of Reproducible Master 16-1, *Installing Strip Flooring*, describe the preparation necessary for wood strip flooring.

10. Explain that strip flooring should be laid at right angles to the floor joists (which is generally the longest dimension of a rectangular room).

11. Referring to Figure 16-5, discuss the size of nails that should be used to nail strip flooring. Mention that the nails should go through the subflooring and into the joists, when possible, to reduce squeaking.

12. Using part *B* of Reproducible Master 16-1, *Installing Strip Flooring,* discuss the nailing technique used for wood strip flooring. Emphasize the importance of making sure the first strip is perfectly aligned with a chalk line to ensure that the remainder of the strips will be square.

13. Describe the procedure that should be used for laying strip flooring around projections such as walls or partitions.

14. Using Reproducible Master 16-2, *Strip Flooring Sequence*, explain how strip flooring is installed throughout several rooms. Using a corner or portion of a room, demonstrate the procedure for installing wood strip flooring. Double-sided "carpet" tape can be used to hold the flooring in position, yet allowing it to be disassembled and used another time.

15. Discuss the procedure used to determine the amount of strip flooring needed for a given area.

16. Describe the steps necessary to prepare a concrete slab for wood strip flooring installation.

17. Explain how the presence of moisture in a concrete slab can be tested.

18. Describe the two procedures that can be used to prepare a slab-on-grade for wood strip flooring installation. Emphasize the importance of preventing moisture from coming into contact with wood strip flooring.

WOOD BLOCK (PARQUET) FLOORING

1. Explain the two methods used to manufacture wood block flooring.

2. Discuss the similarities between preparation of the subfloor and installation of strip flooring and wood block flooring.

3. Using Reproducible Master 16-3, *Laying Out Parquet Flooring*, explain the steps for laying out parquet flooring. Stress the importance of locating the center of the room in both directions and then putting down chalk lines as guides for

laying the first blocks. Explain that the same procedure can be used for laying down resilient tile.

PREFINISHED WOOD FLOORING

1. Cite advantages and disadvantages of using prefinished wood flooring.

UNDERLAYMENT

1. Identify the types of underlayment that are used for vinyl and linoleum. Describe the type of underlayment that might be required on concrete floors.

2. Discuss the preparation necessary before installing hardboard or particleboard as underlayment.

3. Describe the procedure for installing underlayments (hardboard, particleboard, or cement board). Point out the importance, when using hardboard, of leaving spacing at the perimeter for expansion. Point out that hardboard also requires 1/32″ space between panels but that particle board and cement board can be butted at joints.

4. Discuss nail spacing for installing hardboard, particleboard, and cement board. Note also that nailing patterns may be printed on some types of underlayment.

5. Have students list reasons why carpenters prefer plywood underlayment to hardboard or particleboard.

6. Describe the procedure for installing plywood underlayment. Discuss the nailing pattern used.

7. Use a small scale model with 1″ = 1′-0″ or 1 1/2″ = 1′-0″ scale. Form the layout of a room by nailing or screwing plywood "walls" (approximately 8″–12″ high) on three sides of another sheet of plywood representing the floor. Mark the direction and position of the joists, and possibly lay out cabinets or other obstacles. Using 1/8″ plywood or hardboard "panels" cut to scale, demonstrate the layout of the panels. Emphasize the direction of the panels and how they can be positioned to conserve materials.

RESILIENT FLOOR TILE

1. Discuss the steps that should be taken after the underlayment has been laid and before the floor tile is installed. Emphasize the importance of having a smooth surface for the more pliable materials such as vinyl, rubber, and linoleum. Suggest that a layer of felt be applied before linoleum.

2. Describe the procedure for laying out resilient tile. Discuss the purpose of a trial layout to avoid installing border tile that are too narrow.

3. Discuss the procedure for spreading adhesives. Stress the importance of spreading the correct amount of adhesive.

4. Explain the procedure used to set the tile in place. Mention that the tiles should not be slid into position. Emphasize the importance of following the manufacturer's recommendations regarding the rolling of the tile.

5. Describe the "finishing touches" that can be used for resilient tile including a feature strip, border strip, and/or cove base.

SELF-ADHERING TILES

1. Using Figure 16-30, discuss the procedure for installing self-adhering tiles.

SHEET VINYL FLOORING

1. Discuss the procedure for installing sheet vinyl flooring. Explain the methods used to secure the flooring.

UNIT REVIEW

1. Review the unit objectives. Be sure that the students fully understand each objective.

2. Assign *Important Terms, Test Your Knowledge* questions, and *Outside Assignments* on pages 480-481 of the text. Review the answers in class.

3. Assign pages 117-122 of the workbook. Review the answers in class.

EVALUATION

1. Use Procedure Checklist—*Installing Wood Strip Flooring,* to evaluate the techniques that students used to perform these tasks.

2. Use Procedure Checklist—*Installing Resilient Flooring,* to evaluate the techniques that students used to perform these tasks.

3. Use Unit 16 Quiz for in-class evaluation. Correct the quizzes and return them to the students for review.

ANSWERS TO TEST YOUR KNOWLEDGE
TEXT PAGE 481

1. oak
2. 1/2″
3. groove edge
4. spline
5. 24 bd. ft.

6. laminated
7. Because it does not expand and contract like wood.
8. 1/32
9. coated
10. linoleum
11. heated
12. Impervious tile would be preferable since it does not absorb water.

ANSWERS TO WORKBOOK QUESTIONS
PAGES 117-122

1. birch
2. A. Face width.
   B. Tongue.
   C. Undercut.
   D. End groove.
   E. Side groove.
3. C. 3/4″
4. B. clear
5. A. 8″
6. Portable nailer.
7. A. 1/2″
   B. Face nail.
   C. Blind nail.
   D. 50°
8. A. 6″
9. starter strip/extender strip
10. bd. ft.: 614
    Bundles: 26
11. bd. ft.: 489
    Bundles: 11
12. A. 16″
    B. 1 × 2 strip.
    C. Strip flooring.
    D. Polyethylene film.
    E. Polyethylene film.
13. C. Time is saved since prefinished units can be butted directly against baseboards.
14. A. 24
15. A. 6″
    B. 3″
16. telegraphing
17. A. 1/32″
    B. Ring grooved.
18. B. 3′
    D. 5′
19. one-fourth
20. wax
21. D. 12′

ANSWERS TO UNIT 16 QUIZ

1. True.
2. False.
3. False.
4. True.
5. False.
6. False.
7. True.
8. C. plywood
9. B. at the center
10. A. along one sidewall
11. A. 1/4″
12. C
13. B
14. D
15. A

Reproducible Master 16-1 (Part A)
Installing Strip Flooring

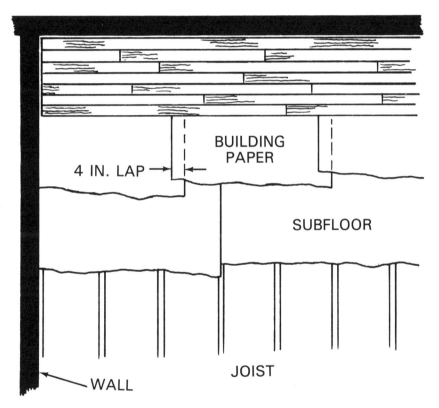

4 IN. LAP

BUILDING PAPER

SUBFLOOR

JOIST

WALL

Installing Strip Flooring

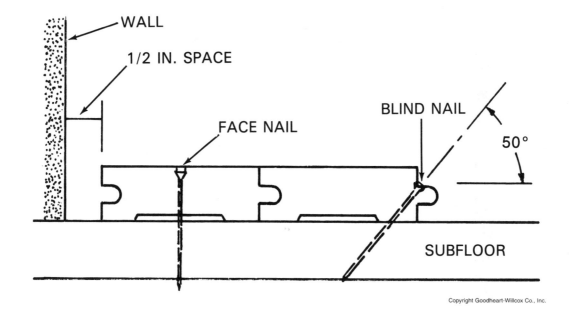

WALL

1/2 IN. SPACE

FACE NAIL

BLIND NAIL

50°

SUBFLOOR

Copyright Goodheart-Willcox Co., Inc.

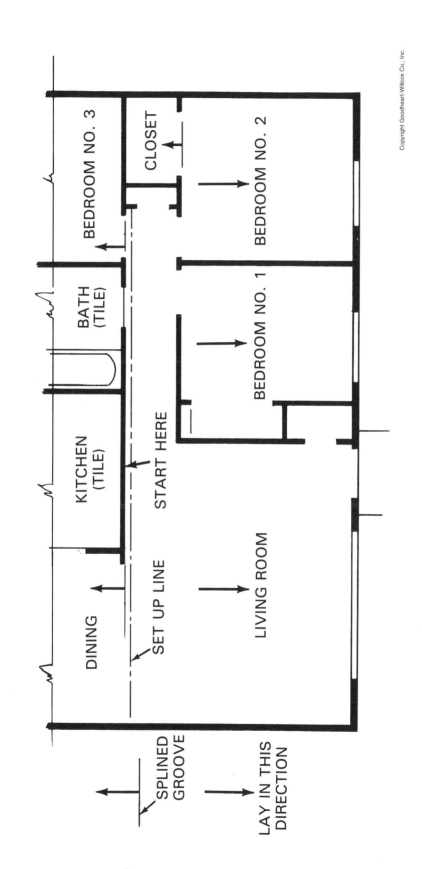

Reproducible Master 16-2

Strip Flooring Sequence

Copyright Goodheart-Willcox Co., Inc.

Reproducible Master 16-3

Laying Out Parquet Flooring Designs

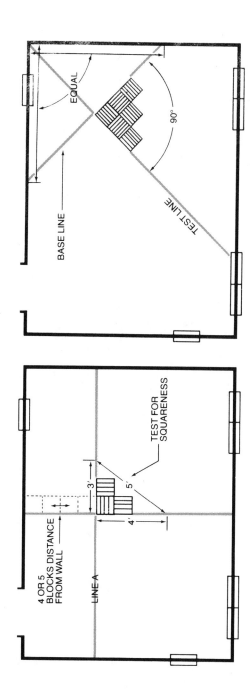

EQUAL

BASE LINE

90°

TEST LINE

4 OR 5 BLOCKS DISTANCE FROM WALL

LINE A

TEST FOR SQUARENESS

3'

4'

5'

Installing Wood Strip Flooring

Name _____ Total _____

❏ Checks subfloor to make sure it is clean and that nail patterns are complete; lays down good-quality building paper, lapping it properly; snaps chalk lines on paper indicating floor joists; applies building paper over duct-work that projects through floor.	5	4	3	2	1	Disregards condition of subfloor; neglects to lay down building paper; does not snap chalk lines; does not apply building paper over ductwork.
❏ Determines best direction for strip flooring; selects appropriate type of nails for flooring installation.	5	4	3	2	1	Does not consider joist direction when determining direction for strip flooring; does not select proper type of flooring nail.
❏ Lays first strip along side wall, allowing 1/2″ space along edge; makes sure the first strip is accurately aligned; properly nails first strip.	5	4	3	2	1	Starts laying wood strips in center of floor; neglects to check alignment of strip.
❏ Blind nails succeeding strips; uses nail set to finish nailing; tightly fits succeeding strips; does not align joints in flooring.	5	4	3	2	1	Face nails all strips; does not blind nail strip flooring; aligns joints in adjacent pieces.

Unit 16 Procedure Checklist
Installing Resilient Flooring

Name _____ Total _____

❑ Installs appropriate underlayment; sweeps and vacuums surface carefully; checks to make sure surfaces are smooth and joints are level; installs base material (if necessary).	5	4	3	2	1	Neglects to install proper underlayment; forgets to sweep and/or vacuum surface; disregards smoothness of surface and joints; neglects to install base material (if needed).
❑ Locates center of tile layout and snaps chalk line; lays out centerline at right angle to the main one and snaps chalk line; makes trial layout of tiles in both directions.	5	4	3	2	1	Neglects to lay out center of room with chalk lines; does not make trial layout.
❑ Removes loose tiles; cleans floor surface; spreads adhesive over one-quarter of room; allows initial set before tile is laid.	5	4	3	2	1	Neglects to clean floor; spreads adhesive over most of room; begins laying tiles as soon as adhesive is spread.
❑ Starts laying tile at center of room, making sure edges align with chalk line; butts adjacent tiles and carefully lays them into place.	5	4	3	2	1	Starts laying tiles along sidewall, disregarding chalk lines; sloppily places tile into position.
❑ Carefully cuts border tiles and sets into position; rolls tile (if necessary).	5	4	3	2	1	Neglects to install border tiles; forgets to roll tile (if required).

Unit 16 Quiz

Finish Flooring

Name _____ Score _____

True-False

Circle T if the answer is True or F if the answer is False.

T F 1. Grading of wood flooring is based on the appearance of the material.

T F 2. Strip flooring should be laid parallel to the floor joists when possible.

T F 3. Conventional subflooring can also be used as underlayment.

T F 4. When laying linoleum, a base layer of felt is recommended.

T F 5. When installing wood strip flooring, carefully align the joints in successive courses.

T F 6. Flexible vinyl flooring is only fastened down around the edges.

T F 7. Wood strip flooring can be installed directly over a concrete floor if the floor is suspended with an air space below.

Multiple Choice

Choose the answer that correctly completes the statement. Write the corresponding letter in the space provided.

_____ 8. In modern construction, _____ is commonly used as a subfloor.

         A. good-quality lumber
         B. gypsum wallboard
         C. plywood
         D. planks

_____ 9. When installing resilient tile, start _____ of the room.

         A. along one sidewall
         B. at the center
         C. at opposite ends
         D. None of the above.

_____ 10. When installing wood strip flooring, start _____ of the room.

         A. along one sidewall
         B. at the center
         C. at opposite ends
         D. None of the above.

_____ 11. Use _____ hardboard for underlayment.

         A. 1/4″
         B. 1/2″
         C. 3/4″
         D. 1″

Identification

Identify the parts of the following section of strip flooring.

_____ 12. Undercut.

_____ 13. Tongue.

_____ 14. Groove.

_____ 15. Face width.

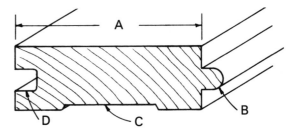

17

Stair Construction

OBJECTIVES

Students will be able to:

- ❏ Recognize the types of stairs.
- ❏ Define basic stair parts and terms.
- ❏ Calculate "rise-run ratio," number and size of risers, and stairwell length.
- ❏ Prepare sketches of the types of stringers.
- ❏ Lay out stringers for a given stair rise and run.
- ❏ List prefabricated stair parts that are commonly available.

INSTRUCTIONAL MATERIALS

Text: Pages 483-500

 Important Terms, page 497

 Test Your Knowledge, page 497

 Outside Assignments, page 498

Workbook: Pages 123-128

Instructor's Manual:

 Reproducible Master 17-1, *Stair Terminology*

 Reproducible Master 17-2, *Stringer Types*

 Unit 17 Quiz

TRADE-RELATED MATH

To calculate the number and size of risers and treads (less the nosing) for given stair run, first divide the total rise by 7. For example, if the total rise for a basement stairway is 8'-4" or 100", the answer will be 14.29. Since there must be a whole number of risers, select the one closest to 14.29 and divide it into the total rise.

$$100'' \div 14 = 7.14 \text{ or } 7\frac{1}{64}$$

Number of risers = 14

Riser height = $7\frac{1}{64}$

In any stairway, the number of treads is always one less than the number of risers. A 10 1/2" tread is

correct for the example, and the total run can be calculated as follows:

Number of treads = 13

$$\text{Total run} = 10\frac{1}{2} \times 13$$

$$= 136.5''$$

$$\text{Total run} = 11'\text{-}4\frac{1}{2}''$$

INSTRUCTIONAL CONCEPTS AND STUDENT LEARNING EXPERIENCES

TYPES OF STAIRS

1. Describe how the construction of stairways has evolved over the years. Explain when stairways are typically installed in a construction project.
2. Identify the two main types of stairs—main stairs and service stairs. Explain that these two types can be open, closed, or a combination of open and closed. Ask students to identify the type of stairs, if any, in their homes.
3. Referring to Figure 17-3, describe the various stairway designs.

PARTS AND TERMS

1. Using Reproducible Master 17-1, *Stair Terminology*, discuss components and terms related to stair construction. Be sure that students can distinguish between unit run and rise and total run and rise.
2. Stress the importance of accurately laying out and constructing any stairway. Describe the types of trimmer and headers that should be used, and where framing anchors should be installed.

STAIR DESIGN

1. Identify the three generally accepted rules for calculating the rise-run or riser-tread ratio. Give the students hypothetical riser and/or tread

dimensions and have them determine the other dimension using the three rules.

2. Discuss problems that may occur if an incorrect riser-tread ratio is used in stair construction. Emphasize the importance of making all risers the same height and all treads the same width when constructing a stairway.

3. Identify the stairway and handrail dimensions that are recommended by some of the model building codes. Have the students refer to their local building code to determine the recommended dimensions for stairways and handrails in your area.

4. Review the stair details included in a set of stock plans or other available residential plans. Point out the dimensions that will be used for the stair construction including the riser height, tread width, stringer size, and rough opening.

STAIR CALCULATIONS

1. Describe the procedure for calculating stair dimensions. Given a hypothetical total run and total rise, have the students calculate number of treads and risers, as well as the dimensions of each.

STRINGER LAYOUT

1. Explain how the riser height is determined using a story pole. Describe how these dimensions are transferred to a stringer and laid out with a framing square. Once again, stress the importance of accuracy.

TREADS AND RISERS

1. Identify the materials that are commonly used for treads and risers of main stairs.

2. Using Figure 17-18, define the term "nosing" and discuss the various designs.

3. Describe the basic stair riser shapes—vertical, sloping, and open.

TYPES OF STRINGERS

1. Using Reproducible Master 17-2, *Stringer Types*, identify the three main types of construction used for stringers, open riser, semihoused, and housed. Also describe or show examples of a cleated stringer. Describe the method of construction used in each.

2. Discuss how a housed stringer stairway is constructed.

WINDER STAIRS

1. Discuss the complaints that are commonly associated with winder stairs.

2. Describe how the wider-tread width is determined for winder stairs.

OPEN STAIRS

1. Using Figure 17-25, identify the parts of an open stair. Stress the importance of anchoring the starting newel post securely to the starting step or carrying it down through the floor and attaching it to the floor frame.

SPIRAL STAIRWAYS

1. Describe the advantages offered by spiral stairways in residential construction.

DISAPPEARING STAIR UNITS

1. Define "disappearing stairs" and how they are used in residential construction.

UNIT REVIEW

1. Review the unit objectives. Be sure that the students fully understand each objective.

2. Assign *Important Terms, Test Your Knowledge* questions, and *Outside Assignments* on pages 497-498 of the text. Review the answers in class.

3. Assign pages 123-128 of the workbook. Review the answers in class.

EVALUATION

1. Use Unit 17 Quiz for in-class evaluation. Correct the quizzes and return them to the students for review.

ANSWERS TO TEST YOUR KNOWLEDGE
TEXT PAGE 497

1. landings
2. 6'-8"
3. 8
4. 24"-25"
5. nosing
6. 4'-7"
7. False.
8. cutout
9. 3/4
10. balusters
11. Instructions incorrect; should read about as follows:

— 5 treads (always 1 less than # of risers) 5 x 11 = 55"

A. Spike housed stringer to wall surface and into wall frame.

B. Set treads and risers into place. Work from top downward using wedges and glue.

ANSWERS TO WORKBOOK QUESTIONS
PAGES 123-128

1. underlayment
2. A. Open.
   B. Open.
   C. Closed.
   D. Combination.
3. A. 4'
4. A. Tread.
   B. Riser.
   C. Unit run.
   D. Unit rise.
   E. Total run.
   F. Total rise.
   G. Headroom.
5. C. 30° to 35°
6. C. 24 to 25
7. B. 11 1/2"-12 1/2"
8. B. 2'-8"
9. A. 30/36"
   B. 34/42"
10. No. of Risers: 15
    Riser Height: 7 1/8"
    No. of Treads: 14
    Tread Width: 10 3/4"
11. 12'-6 1/2"
12. story pole
13. C. shortening the bottom riser by an amount equal to tread thickness

14. A. 1 1/2"
    B. decreased
15. A. Vertical.
    B. Sloping.
    C. Open.
16. B. one-third
17. semi-housed
18. D. 3/4"
19. outside
20. A. Closed stringer.
    B. Newel.
    C. Handrail.
    D. Balusters.
    E. Open stringer.
21. C. nail set
22. C. newel

ANSWERS TO UNIT 17 QUIZ

1. E
2. F
3. C
4. D
5. B
6. H
7. J
8. I
9. G
10. A
11. B. headroom
12. C. 17"-18"
13. B. 3'-0"
14. D. stringers
15. A. balustrade

Reproducible Master 17-1
Stair Terminology

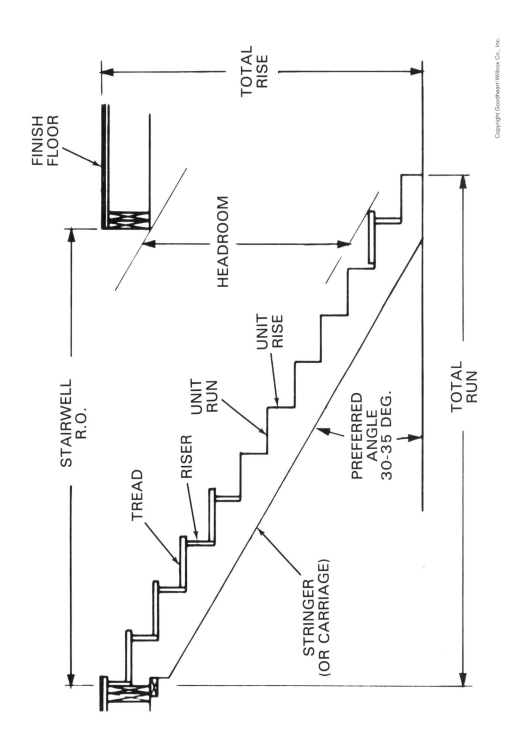

TOTAL RISE

FINISH FLOOR

HEADROOM

STAIRWELL R.O.

TREAD

RISER

UNIT RUN

UNIT RISE

STRINGER (OR CARRIAGE)

PREFERRED ANGLE 30-35 DEG.

TOTAL RUN

Reproducible Master 17-2
Stringer Types

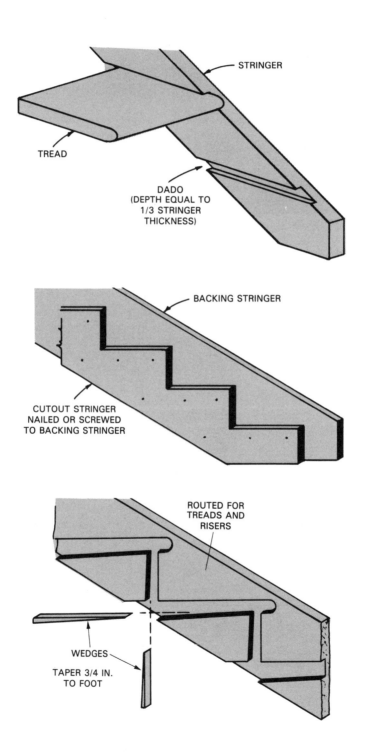

STRINGER

TREAD

DADO
(DEPTH EQUAL TO
1/3 STRINGER
THICKNESS)

BACKING STRINGER

CUTOUT STRINGER
NAILED OR SCREWED
TO BACKING STRINGER

ROUTED FOR
TREADS AND
RISERS

WEDGES

TAPER 3/4 IN.
TO FOOT

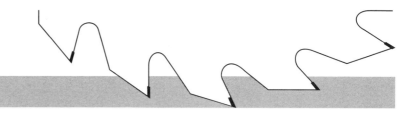

Unit 17 Quiz
Stair Construction

Name _____ Score _____

Identification

Identify the basic stair parts and related terms.

_____ 1. Unit run.

_____ 2. Unit rise.

_____ 3. Tread.

_____ 4. Riser.

_____ 5. Finish floor.

_____ 6. Total rise.

_____ 7. Total run.

_____ 8. Stringer.

_____ 9. Headroom.

_____ 10. Stairwell rough opening.

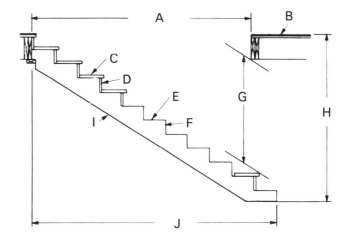

Multiple Choice

Choose the answer that correctly completes the statement. Write the corresponding letter in the space provided.

_____ 11. The vertical space above the stair is the _____.

    A. total rise
    B. headroom
    C. unit rise
    D. None of the above.

_____ 12. The sum of one riser and one tread should equal _____ .

    A. 9"-10"
    B. 13"-14"
    C. 17"-18"
    D. 21"-22"

_____ 13. A minimum width of _____ is generally recommended for main stairs.

    A. 2'-6"
    B. 3'-0"
    C. 3'-6"
    D. 4'-0"

_____ 14. Treads and risers are supported by the _____.

A. finish floor
B. tail joists
C. headers
D. stringers

_____ 15. The principal members of a(n) _____ are the newel, baluster, and handrail.

A. balustrade
B. open stringer
C. closed stringer
D. None of the above.

18

Doors and Interior Trim

OBJECTIVES

Students will be able to:

- ❏ Describe how door frames and casings are installed.
- ❏ Explain the difference between panel- and flush-type doors.
- ❏ List the steps for hanging a door.
- ❏ Name lock parts and describe typical installation procedures.
- ❏ Compare pocket- and bypass-types of sliding doors.
- ❏ Outline the order in which window trim members should be applied.
- ❏ Cut, fit, and nail baseboard trim.

INSTRUCTIONAL MATERIALS

Text: Pages 501-526
    Important Terms, page 525
    Test Your Knowledge, page 525
    Outside Assignments, page 525
Workbook: Pages 129-136
Instructor's Manual:
    Reproducible Master 18-1, *Typical Moldings, Parts A* and *B*
    Reproducible Master 18-2, *Casing Interior Doors*
    Reproducible Master 18-3, *Installing Baseboard and Base Shoe, Parts A* and *B*
    Procedure Checklist—*Installing Baseboards and Base Shoe*
    Unit 18 Quiz

TRADE-RELATED MATH

Base trim is determined by calculating the perimeter of the room and adding 5% waste. The perimeter of a room is determined by adding the overall dimensions. When the room is square or rectangular, twice the width can be added to twice the length of the room. For example, if the room dimensions are $8' \times 14'$, determine the perimeter.

$$Perimeter = 2W + 2L$$
$$= (2 \times 8') + (2 \times 14')$$
$$Perimeter = 44$$

Five percent waste is then calculated as follows:

$$Total\ linear\ feet = perimeter + (perimeter \times 5\%)$$
$$= 44' + (44' \times .05)$$
$$= 44' + 2.2'$$
$$Total\ linear\ feet = 46.2$$

INSTRUCTIONAL CONCEPTS AND STUDENT LEARNING EXPERIENCES

MOLDINGS

1. Describe the purpose of moldings, and identify the moldings shown in parts *A* and *B* of Reproducible Master 18-1, *Typical Moldings.*

INTERIOR DOOR FRAMES

1. Identify the components of a doorframe. Discuss the types of materials used for doorframes. Explain why the back sides of jambs are usually kerfed.
2. Obtain an adjustable doorjamb and describe how it can be adjusted to accommodate different wall thicknesses.

INSTALLING DOOR FRAMES

1. Describe the installation procedure used for interior door frames. Explain the purpose of the spreader that is commonly used between the side jambs. Demonstrate the techniques used for installing door frames.
2. Using Reproducible Master 18-2, *Casing Interior Doors,* explain the purpose of a door casing. Have the students note that the casing covers the blocking, and is attached to the jamb and wall surface.
3. Describe how the side pieces are laid out and cut. List the types of nails used to attach them to the jamb and studs, and how they are attached.

PANEL DOORS

1. Identify the two general types of doors—panel and flush. Cite characteristics of each.
2. Identify the parts of a panel door, and list some of the materials used for these parts.

FLUSH DOORS

1. Describe the construction of flush doors by referring to Figure 18-15. Identify the types of materials used in their construction.
2. List the standard sizes for interior and exterior doors used in residential construction.
3. Identify the interior and exterior door designs shown in Figures 18-18 and 18-19.

DOOR INSTALLATION

1. List the two pieces of information regarding doors that can be determined from the architectural drawings and door schedule.
2. Cite reasons why doors should not be cut to fit smaller openings.
3. Discuss the storage methods that should be used for doors. Have the students explain why doors should be conditioned a few days before installation.
4. Identify the recommended clearances for interior doors. Describe how doors can be trimmed to exact size. Explain the purpose of beveling the edge on the lock side, and rounding all edges of the door.
5. Demonstrate the use of a door-and-jamb template if one is available. Emphasize the importance of accuracy when cutting gains.
6. Discuss the procedure used to mount the hinges on a door and jamb. Demonstrate this procedure as well.
7. Explain the purpose of a doorstop. Describe how one is installed.
8. Using Figure 18-27, identify the types of lock sets that are used in residential construction. Have the students rate the locks based on security and ease of installation.
9. Describe how to determine the "hand of the door." Determine the hand of the door for doors in your classroom or on working drawings.
10. Discuss the purpose of a deadbolt.

LOCK INSTALLATION

1. Describe how locks are installed. Mention that the locks are commonly installed 36″ to 38″ from the floor.
2. Identify the various tools that can be used to drill

the holes for locks and make the mortise for the faceplate.

THRESHOLDS AND DOOR BOTTOMS

1. Explain the purpose of a threshold, and list the materials that are used to make them.

PREHUNG DOOR UNIT

1. Identify the advantages of using prehung door units.
2. Using Figure 18-38, describe the procedure for installing prehung doors.

SLIDING DOORS (POCKET-TYPE)

1. Identify the primary advantage of pocket-type bypass doors. Describe the installation procedure required for this type of bypass door.

SLIDING DOORS (BYPASS-TYPE)

1. Discuss the procedure for installing bypass-type sliding doors.
2. Identify the hardware components used for bypass doors.

FOLDING DOORS (BIFOLD)

1. Using Figure 18-46, describe the hardware components that are commonly used for folding doors.

MULTIPANEL FOLDING DOORS

1. Have the students list applications for multipanel folding doors. In addition, have them identify advantages of this type of door.

WINDOW TRIM

1. Using Figure 18-50, identify the trim members used for a double-hung window.
2. Describe the procedure for installing trim around a window.
3. Demonstrate the procedure for cutting a returned end. Discuss the purpose of a returned end.

BASEBOARD AND BASE SHOE

1. Discuss the installation of the baseboard and base shoe using part *A* of Reproducible Master 18-3, *Installing Baseboard and Base Shoe*. Mention that even though the baseboard and base shoe are fitted at the same time, the base shoe usually is not nailed into position until after the floor surface finishes are applied.
2. Using part *B* of Reproducible Master 18-3, *Installing Baseboard and Base Shoe,* identify

where coped joints and miter joints will be used. Demonstrate the procedure for cutting coped and mitered joints for baseboard trim members.

UNIT REVIEW

1. Review the unit objectives. Be sure that the students fully understand each objective.
2. Assign *Important Terms, Test Your Knowledge* questions, and *Outside Assignments* on page 525 of the text. Review the answers in class.
3. Assign pages 129-136 of the workbook. Review the answers in class.

EVALUATION

1. Use Procedure Checklist—*Installing Baseboards and Base Shoe,* to evaluate the techniques that students used to perform these tasks.
2. Use Unit 18 Quiz for in-class evaluation. Correct the quizzes and return them to the students for review.

ANSWERS TO TEST YOUR KNOWLEDGE
TEXT PAGE 525

1. 5 1/4
2. lugs
3. stops
4. casing
5. A plinth block is a decorative corner trim used mainly on the trim around windows and doors. It eliminates mitered corners.
6. stiles
7. skins
8. ladder
9. 2'-6", 6'-8"
10. 1/16"
11. LHR
12. tubular
13. pocket type
14. side casing
15. apron
16. coped

ANSWERS TO WORKBOOK QUESTIONS
PAGES 129-136

1. A. Quarter round.
   B. Cove.
   C. Base shoe.
   D. Stop.
   E. Mullion casing.
   F. Baseboard.
   G. Bed Mold
   H. Casing.

2. C. 5 1/4"
3. 6'-9"
4. C. 8d
5. A. 1 3/4"
   B. 1 3/8"
6. A. 3/16"
   B. Side jamb.
   C. Blocking.
   D. Casing.
7. D. Use 8d casing or finish nails to secure the casing to the jamb and space them about 10" apart.
8. A. horizontal
   B. vertical
9. A. Core.
   B. Stile.
   C. Rail.
   D. Core.
   E. Plywood face/skin.
   F. Lock block.
10. B. 40%
11. A. 3/32"
    B. 1/16"
    C. 1/16"
    D. 5/8"
12. D. 3 1/2"
13. Top hinge: B. 7"
    Bottom hinge: E. 11"
14. B. 1/16"
15. A. Cylindrical.
    B. Tubular.
    C. Unit.
16. deadbolt/deadlock
17. A. LHR.
    B. HR.
    C. LH.
    D. RHR.
18. 36"/38"
19. A. templates
20. A. Knob.
    B. Front plate/faceplate.
    C. Bolt.
    D. Rose.
    E. Cylinder.
21. pocket door
22. bifold
23. bypass
24. D. stack
25. A. Side casing.
    B. Mullion.
    C. Head casing.
    D. Stool.

E. Apron.

26. A. apron
    B. side casing
27. A. Miter.
    B. Coped.
    C. Mitered-lap/scarf.
    D. Butt.

ANSWERS TO UNIT 18 QUIZ

1. B. 1 3/4″
2. C. 36″-38″
3. D. left-hand
4. B. baseboard

5. E
6. B
7. A
8. F
9. D
10. C
11. D
12. C
13. G
14. A
15. B
16. E
17. F

Reproducible Master 18-1 (Part A)
Typical Moldings

CEILING TREATMENTS

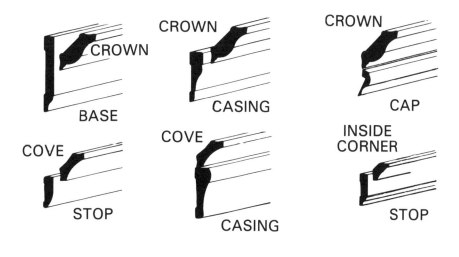

CHAIR RAILS

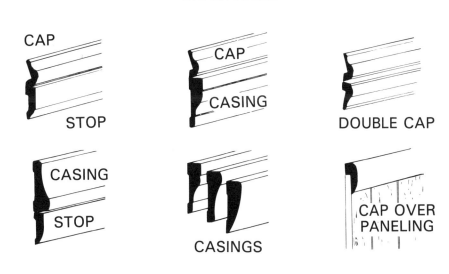

FLOOR AND WALL CORNERS

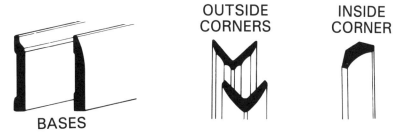

Reproducible Master 18-1 (Part B)
Typical Moldings

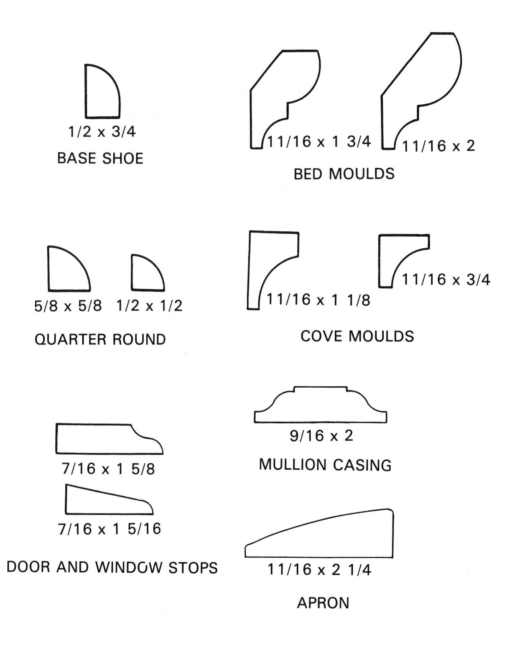

1/2 x 3/4
BASE SHOE

11/16 x 1 3/4   11/16 x 2
BED MOULDS

5/8 x 5/8   1/2 x 1/2
QUARTER ROUND

11/16 x 1 1/8   11/16 x 3/4
COVE MOULDS

7/16 x 1 5/8

9/16 x 2
MULLION CASING

7/16 x 1 5/16
DOOR AND WINDOW STOPS

11/16 x 2 1/4
APRON

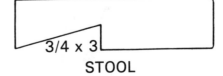

3/4 x 3
STOOL

Reproducible Master 18-2
Casing Interior Doors

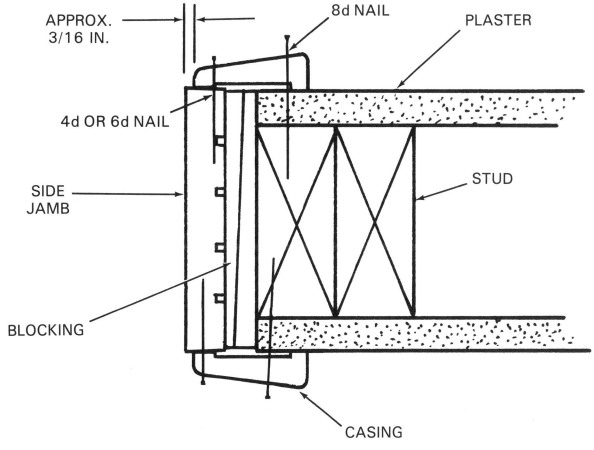

APPROX.
3/16 IN.

8d NAIL

PLASTER

4d OR 6d NAIL

SIDE
JAMB

STUD

BLOCKING

CASING

Installing Baseboard and Base Shoe

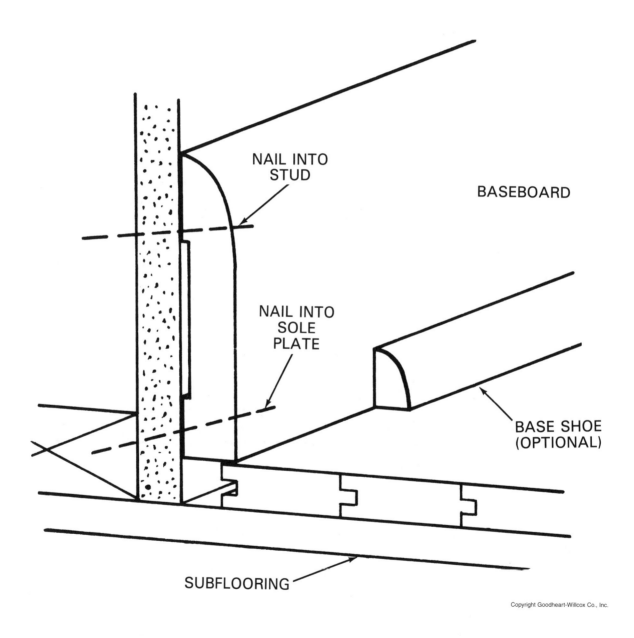

NAIL INTO
STUD

BASEBOARD

NAIL INTO
SOLE
PLATE

BASE SHOE
(OPTIONAL)

SUBFLOORING

Reproducible Master 18-3 (Part B)

Installing Baseboard and Base Shoe

MITER
JOINT

COPED
JOINT

Unit 18 Procedure Checklist
Installing Baseboard and Base Shoe

Name _____ Total _____

❏ Determines where baseboard material will be located; sorts pieces to reduce amount of waste.	5	4	3	2	1	Cannot determine where baseboard material is to be located; disregards lengths of baseboard materials when determining where cuts are to be made.
❏ Locates positions of wall studs; marks wall for nailing of baseboard.	5	4	3	2	1	Forgets to mark positions of wall studs.
❏ Carefully cuts and fits first piece of baseboard; uses scarf joints when connecting lengths of baseboard.	5	4	3	2	1	Miscalculates length of first piece of baseboard; uses butt joints when connecting lengths of baseboard.
❏ Creates coped joints for inside corners; uses miter joints on exterior corners; properly nails baseboard into position.	5	4	3	2	1	Uses butt joints for inside and exterior corners; misses studs when nailing baseboard.
❏ Installs base shoe after floor finish has been applied.	5	4	3	2	1	Installs base shoe immediately after baseboard, disregarding whether floor finish is to be applied.

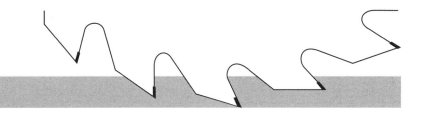

Unit 18 Quiz
Doors and Interior Trim

Name _____ Score _____

Multiple Choice

Choose the answer that correctly completes the statement. Write the corresponding letter in the space provided.

_____ 1. The standard thickness for exterior doors is _____ .

A. 1 1/2″
B. 1 3/4″
C. 2″
D. 2 1/4″

_____ 2. Locksets are usually installed _____ from the floor.

A. 24″-26″
B. 30″-32″
C. 36″-38″
D. 42″-44″

_____ 3. A door that is hinged on the left and opens inward is referred to as a _____ door.

A. right-hand reverse
B. left-hand reverse
C. right-hand
D. left-hand

_____ 4. The _____ covers the joint between the wall surface and finish flooring.

A. base shoe
B. baseboard
C. crown molding
D. stool

Identification

Identify the parts of the panel door.

_____ 5. Stile.

_____ 6. Mullion.

_____ 7. Top rail.

_____ 8. Bottom rail.

_____ 9. Lock rail.

_____ 10. Panel.

Identify the various types of moldings.

_____ 11. Cove mold.

_____ 12. Quarter round.

_____ 13. Stool.

_____ 14. Base shoe.

_____ 15. Bed mold.

_____ 16. Mullion casing.

_____ 17. Apron.

19

Cabinetmaking

OBJECTIVES

Students will be able to:

- ❑ Select prefabricated cabinets for a specific floor plan.
- ❑ Install prefabricated base and wall cabinets.
- ❑ Compare the common alternative procedures for building cabinets on the job.
- ❑ Lay out and frame a cabinet from drawings.
- ❑ Describe the three types of drawer guides.
- ❑ List the steps in cutting and assembling drawers.
- ❑ Discuss material choices for cabinet shelves and doors.
- ❑ Explain how to install a plastic laminate surface.

INSTRUCTIONAL MATERIALS

Text: Pages 527-556

Important Terms, page 555
Test Your Knowledge, page 555
Outside Assignments, page 556

Workbook: Pages 137-144

Instructor's Manual:

Reproducible Master 19-1, *Standard Cabinet Dimensions*
Reproducible Master 19-2, *Cabinet Styles*
Reproducible Master 19-3, *Installing Factory-Built Cabinets*
Procedure Checklist—*Installing Factory-Built Cabinets*
Unit 19 Quiz

TRADE-RELATED MATH

Many times a carpenter will be required to find the middle of a dimension or divide fractional amounts by 2. To divide a fraction by 2, simply multiply the bottom number (denominator) by 2. For example:

$$\frac{1}{2} \div 2 = \frac{1}{2 \times 2} = \frac{1}{4}$$

$$\frac{1}{2} \div 2 = \frac{1}{4}$$

A mixed number (number containing both a whole number and a fraction) can be divided by 2 using three steps. First, divide the whole number by 2. The answer will always be a whole number (2, 4, 6, etc.) or a whole number and one-half (2 1/2, 3 1/2, 5 1/2, etc.). Next, divide the fraction by 2 using the method shown in the previous example. Finally, add the two parts together. (You may need to find a common denominator before adding, however.)

$$8\frac{1}{2} \div 2$$

$$8 \div 2 = 4$$

$$\frac{1}{2} \div 2 = \frac{1}{4}$$

$$4 + \frac{1}{4} = 4\frac{1}{4}$$

$$8\frac{1}{2} \div 2 = 4\frac{1}{4}$$

INSTRUCTIONAL CONCEPTS AND STUDENT LEARNING EXPERIENCES

DRAWINGS FOR CABINETWORK

1. List the drawings that commonly show information related to cabinetwork. Using a stock plan that includes cabinet layout and details (or other comparable set of plans), show the students these drawings and describe the primary dimensions that will be needed in the construction.

STANDARD SIZES

1. Explain that the overall heights and other dimensions of built-in cabinets are standardized. Using Reproducible Master 19-1, *Standard Cabinet Dimensions*, identify the common dimensions of cabinetwork. List other pieces of furniture and appliances that usually have standard dimensions.

Factory Built Cabinets

1. Point out that most builders today use factory-built cabinets, although occasionally they may be built on site. Discuss the three forms in which actory-built cabinets may be delivered: disassembled, assembled but not finished, and assembled and finished. Cover the advantages and disadvantages of each.
2. Using Reproducible Master 19-2, *Cabinet Styles*, review the different styles in which cabinets are made: frame with cover (or face-frame) and frameless.
3. Ask your students to secure literature and information about standard cabinet units from local building supply stores. Have them compare features and dimensions of the various cabinets to see if there are similarities among cabinets built by different manufacturers.
4. Identify and discuss the types of natural or manufactured materials used in factory-built cabinets. Compare qualities of various materials.
5. Discuss joinery and other fastening techniques and compare quality. Recognizing the marks of quality in construction is important.
6. Using Reproducible Master 19-3, *Installing Factory-Built Cabinets*, review the standard procedure for installing cabinets. Note that some builders prefer to install the wall cabinets before the base cabinets. Discuss the advantages of this practice.
7. If a set of base and wall cabinets are available, demonstrate the correct procedure for installation. Involve your students in this task.

BUILDING CABINETS

1. Discuss the approaches used for building cabinets on the job. List advantages and disadvantages for each approach.
2. Describe how a master layout is made for a cabinet, and discuss its use. Group the students in pairs and provide each with a set of drawings for cabinetwork. Have each pair prepare a master layout for cabinets. When all groups have completed, rotate the groups and have them check another group's layout. Have the students mark any discrepancies with a pen or marker.
3. Describe the basic framing of base cabinets. Using a scale model, demonstrate how the base cabinets are framed in.
4. Identify the materials that are commonly used for cabinet facings. Add the facing materials to the scale model of the cabinet.

5. Identify different styles of drawer guides that may be used (Figure 19-36). Emphasize that many cabinetmakers use manufactured drawer guides.
6. Describe the two general types of drawers. Show the students the construction techniques used for manufactured drawers, citing good and poor techniques.
7. Review the use of the table saw, emphasizing the safety rules. Demonstrate the construction of lip and flush doors and point out the main differences in construction. Stress the importance of squaring the drawer up before gluing or nailing it together.
8. Have the pairs of students construct one drawer for their cabinet. Make sure that the drawer fits their cabinet.
9. Identify the different means of supporting shelves in cabinetwork. List the advantages and disadvantages of each type.
10. Discuss the maximum spacing that should be used for shelves. Explain that the function of the cabinet dictates what type of spacing and shelf support is used.
11. Identify the types of doors—flush, overlay, and lip—that are used in cabinet construction. Have the students discuss the major differences.
12. Demonstrate the construction techniques used for flush, lip, and overlay doors.
13. Have the pairs of students construct a door for their cabinet.
14. Identify the types of hinges that are used for flush cabinet doors (shown in Figure 19-49).
15. Discuss applications for sliding doors in cabinetry. Describe the different types of tracks or door designs that are used for sliding doors.
16. Identify the types of materials that are used for cabinet tops. Emphasize the importance of using a backing material that is dimensionally stable and suitable for lamination.
17. Demonstrate how laminates are laid out and cut. Stress the proper cutting side to avoid splintering and cracking.
18. Emphasize the precautions that should be taken when using contact cement for bonding laminates to a backing material. Using a cabinet top sized for your scale model, demonstrate the preparation of the base material. When the base has been properly prepared, show the students how to apply the contact cement, and then apply the plastic laminate.
19. When the adhesive has properly set, demonstrate

the procedure for trimming the edges using a laminate trimmer or router.

20. Have each pair of students construct a top for their cabinet.

21. Identify various types of hardware that can be used for cabinetry.

OTHER BUILT-IN UNITS

1. Review advantages and disadvantages of built-in cabinets. Ask students to list the built-in cabinets they have in their homes, their location, and their purpose.

SEQUENCE OF INTERIOR FINISH

1. Discuss the sequence of events that occur in finishing the interior of a house. Stress that good communication with other tradespeople is vital to avoid bottlenecks and interference with other workers.

UNIT REVIEW

1. Review the unit objectives. Be sure that the students fully understand each objective.

2. Assign *Important Terms, Test Your Knowledge* questions, and *Outside Assignments* on pages 555-556 of the text. Review the answers in class.

3. Assign pages 137-144 of the workbook. Review the answers in class.

EVALUATION

1. Use Procedure Checklist—*Installing Factory-Built Cabinets,* to evaluate the techniques that students used to perform these tasks.

2. Use Unit 19 Quiz for in-class evaluation. Correct the quizzes and return them to the students for review.

ANSWERS TO TEST YOUR KNOWLEDGE
TEXT PAGE 555

1. A. Built on the job.
   B. Custom-built in local cabinet shop.
   C. Mass-produced from cabinet factory.
2. architectural plans
3. frameless
4. False (holes are 5 mm).
5. C. Checks walls and floors for high spots.
6. story stick
7. False. Some installers prefer to install wall cabinets first.
8. 36
9. master layout
10. after

11. stiles
12. before
13. center guides
14. 3/8, 1/2
15. 42
16. overlay
17. backing sheet
18. short
19. above

ANSWERS TO WORKBOOK QUESTIONS
PAGES 137-144

1. attached
2. D. carpenter
3. B. 24″
4. C. assembled but not finished
5. A. Magnetic
   B. Friction.
   C. Ball/bullet.
6. A. Face frame.
   B. Frameless.
7. A. 30″
   B. 7′-0″
   C. 36″
   D. 6′-0″
   E. 31″
   F. 5′-4″
   G. 2′-0″
8. No. 3012: 30″ × 12″ × 12 1/2″
   No. 27303: 27″ × 30″ × 12 1/2″
9. toggle
10. C. Fit the counter top into position and attach it to the cabinets before cabinets are fastened to wall.
11. shims, blocking
12. master layout
13. A. Faceplate/face frame.
    B. End panel.
    C. Web frame.
    D. Dado joint.
    E. Rabbet joint.
14. rails
15. B. End panels are installed after bottom is attached to the base.
16. C. Marked piece should not be used to lay out duplicate parts.
17. C. 3/8″
18. A. Corner.
    B. Center.
    C. Side.
19. A. kicker
20. D. front and sides

21. D. top and sides
22. A. front
23. See Figure 19-40 in text. (One point for each correct drawing.)
24. before
25. D. 42″
26. A. Flush.
    B. Overlay.
    C. Lip.
27. C. gain
28. B. raising the door and pulling the lower edge outward
29. B. 1/16″
30. B. contact bond cement
31. C. 4 sq. ft.
32. B. 20°
33. above
34. damage

35. A. 2
    B. 4
    C. 5
    D. 1
    E. 3

ANSWERS TO UNIT 19 QUIZ

1. True.
2. False.
3. True.
4. True.
5. C. 36″
6. B. stiles
7. A. 42″
8. B. width
9. C. lipped
10. C. Fir plywood.
11. D. All of the above.

Reproducible Master 19-1
Standard Cabinet Dimensions

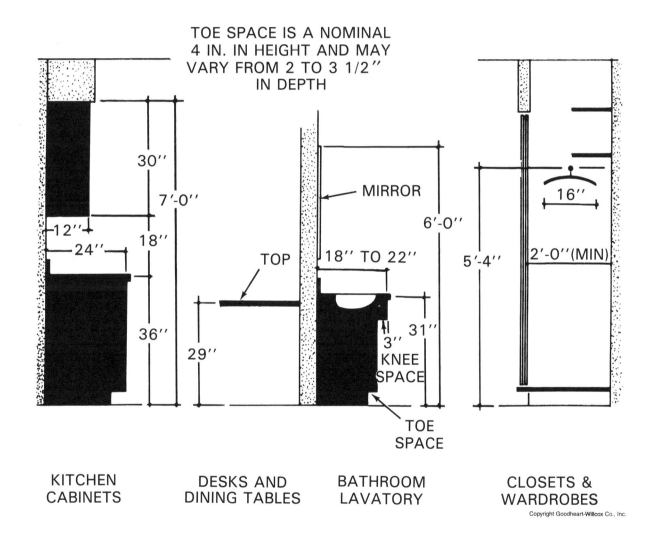

TOE SPACE IS A NOMINAL
4 IN. IN HEIGHT AND MAY
VARY FROM 2 TO 3 1/2″
IN DEPTH

30″

7'-0″

12″

24″

18″

36″

29″

TOP

MIRROR

18″ TO 22″

6'-0″

5'-4″

16″

2'-0″(MIN)

3″
KNEE
SPACE

31″

TOE
SPACE

KITCHEN
CABINETS

DESKS AND
DINING TABLES

BATHROOM
LAVATORY

CLOSETS &
WARDROBES

Reproducible Master 19-2
Cabinet Styles

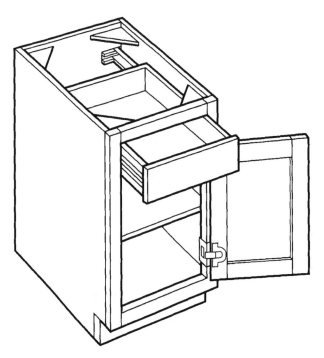

FACE-FRAME
(frame with cover)

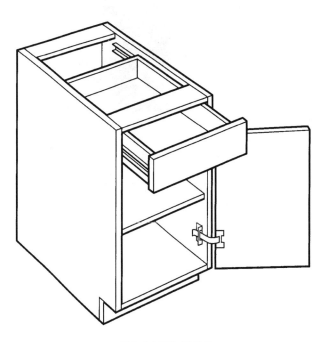

FRAMELESS

Reproducible Master 19-3 **Installing Factory-Built Cabinets**

1.
Locate the position of all wall studs where cabinets are to hang by tapping with a hammer. Mark their position where the marks can easily be seen when the cabinets are in position.

2.
Find the highest point on the floor with a level. This is important for both base and wall cabinet installation later. Remove the baseboard from all walls where cabinets are to be installed. This will allow them to go flush against the walls.

3.
Start the installation with a corner or end unit. Slide it into place then continue to slide the other base cabinets into the proper position.

4.
When all base cabinets are in position, fasten the cabinets together. This is done by drilling a 1/4'' diameter hole through the face frames and using the 3'' screws and T-nuts provided. To get maximum holding power from the screw, one hole should be close to the top of the end stile and one should be close to the bottom.

5.
Check the position of each cabinet with a spirit level, going from the front of the cabinet to the back of the cabinet. Next shim between the cabinet and the wall for a perfect base cabinet installation.

6.
Starting at the high point in the floor, level the leading edges of the cabinets. Continue to shim between the cabinets and the floor until all the base cabinets have been brought to level.

7.
After the cabinets have been leveled, both front to back and across the front, fasten the cabinets to the wall at the stud locations. This is done by drilling a 3/32'' diameter hole 2 1/4'' deep through both the hanging strips for the 2 1/2'' x 8 screws that are provided.

8.
Fit the counter top into position and attach it to the base cabinets by predrilling and screwing through the front corner blocks into the top. Use caution not to drill through the top. Cover the counter top for protection while the wall cabinets are being installed.

9.
Position the bottom of the 30'' wall cabinets 19'' from the top of the base cabinet, unless the cabinets are to be installed against a soffit. A brace can be made to help hold the wall cabinets in place while they are being fastened. Start the wall cabinets installation with a corner or end cabinet. Use care in getting this cabinet installed plumb and level.

10.
Temporarily secure the adjoining wall cabinets so that leveling may be done without removing them. Drill through the end stiles of the cabinets and fasten them together as was done with the base cabinets.

11.
Use a spirit level to check the horizontal surfaces. Shim between the cabinet and the wall until the cabinet is level. This is necessary if doors are to fit properly.

12.
Check the perpendicular surface of each frame at the front. When the cabinets are level, both front to back and across the front, permanently attach the cabinets to the wall. This is done by predrilling a 3/32'' diameter hole 2 1/4'' deep through the hanging strip inside the top and below the bottom of the cabinets at the stud location. Enough Number 8 screws should be used to fasten the cabinets securely to the wall.

Unit 19 Procedure Checklist
Installing Factory-Built Cabinets

●

Name _____ Total _____

❑ Locates wall studs where cabinets are to be placed and marks their position; finds highest point on the floor with a level; removes base board where base cabinets are to be placed.	5	4	3	2	1	Neglects to find studs or highest point on floor; forgets to remove baseboard where base cabinets will be installed, or removes baseboard where cabinets are not to be installed.
❑ Starts base cabinet installation in corner; slides remainder of base cabinets against adjacent cabinet; fastens all cabinets together.	5	4	3	2	1	Starts base cabinet installation in middle; neglects to fasten cabinets together.
❑ Checks position of cabinets with level; shims between cabinets and all; levels leading edges of cabinets, shimming between the cabinets and floor.	5	4	3	2	1	Neglects to check position of cabinets.
❑ Fastens cabinets to wall at stud locations.	5	4	3	2	1	Neglects to securely fasten cabinets to wall.
❑ Fits countertop to base cabinets.	5	4	3	2	1	Neglects to fit countertop to base cabinets.
❑ Positions wall cabinets, and uses brace to hold in position; starts installation with corner or end cabinet.	5	4	3	2	1	Tries to hold wall cabinets in position while marking position; starts installation in middle of wall.
❑ Temporarily secures adjacent cabinets; checks level of wall cabinets, shimming between the cabinet and wall; checks the plumb of the cabinet using a level.	5	4	3	2	1	Neglects to check level and plumb of cabinets.

●

●

Unit 19 Quiz
Cabinetmaking

Name _____ Score _____

True-False

Circle T if the answer is True or F if the answer is False.

T F 1. Horizontal members of a face frame are called rails.

T F 2. A kicker is placed along the bottom of a base cabinet to provide clearance for a person's toes.

T F 3. Plywood should not be used for drawer sides and backs.

T F 4. When using a table saw to cut plastic laminate, the decorative side should face upward.

Multiple Choice

Choose the answer that correctly completes the statement. Write the corresponding letter in the space provided.

_____ 5. Base kitchen cabinets are usually _____ high.
   A. 30″
   B. 33″
   C. 36″
   D. 39″

_____ 6. The vertical members of a face frame are called _____.
   A. rails
   B. stiles
   C. facing strips
   D. None of the above.

_____ 7. Standard 3/4″ shelving should be supported every _____ or closer.
   A. 42″
   B. 48″
   C. 54″
   D. 60″

_____ 8. When specifying the size of the opening of a cabinet door, always list the _____ first.
   A. thickness
   B. width
   C. height
   D. None of the above.

_____ 9. A(n) _____ cabinet door is rabbetted along all edges so that part of the door is inside the door frame.
   A. flush
   B. overlay
   C. lipped
   D. All of the above.

_____ 10. Which of the following is *not* satisfactory as a base for plastic laminates?

   A. Particleboard.
   B. Hardboard.
   C. Fir plywood.
   D. Waferboard.

_____ 11. Which of the following can be used to apply contact cement to a base material?

   A. Brush.
   B. Roller.
   C. Spreader.
   D. All of the above.

20

Painting, Finishing, and Decorating

OBJECTIVES

Students will be able to:
- ❏ Cite safety rules that apply to painting and finishing
- ❏ List tools and equipment and demonstrate their use.
- ❏ Select proper materials for various painting, finishing and decorating jobs.
- ❏ Prepare exterior and interior surfaces for painting.
- ❏ Explain proper procedures for painting, finishing and wallpaper hanging

INSTRUCTIONAL MATERIALS

Text: Pages 557-577
  Important Terms, page 576
  Test Your Knowledge, page 576
  Outside Assignments, page 577
Workbook: Pages 145-148
Instructor's Manual:
  Reproducible Master 20-1, *Safety Rules for Painting and Finishing*
  Reproducible Master 20-2, *Parts of a Paint Brush*
  Reproducible Master 20-3, *Types of Brushes*
  Reproducible Master 20-4, *Parts of a Spray Gun*
  Unit 20 Quiz
  Section 4 Exam

TRADE-RELATED MATH

Simple multiplication and division are used to determine the amount of paint needed to cover areas to be finished. In the first step, the dimensions of the wall or ceiling are measured; then the two dimensions are multiplied to determine the square footage. Finally, long division is used to determine the amount of paint required.

For example:

How much paint will be needed to paint a wall that is 8′ high and 14′ long if the coverage is 400 Sq. ft./gal?

wall height × wall length = sq. ft.

$14' \times 8' = 112$ sq. ft.

$\dfrac{\text{Number of sq. ft.}}{\text{Coverage/gal}} = $ Quantity of paint needed

$\dfrac{112 \text{ sq. ft.}}{400 \text{ sq. ft./gal}} = 0.28$ gal.

INSTRUCTIONAL CONCEPTS AND STUDENT LEARNING EXPERIENCES

SAFETY WITH PAINTS AND FINISHES

1. Review the safety rules at the beginning of this unit and emphasize the need to take precautions when working with coatings. Distribute Reproducible Master 20-1, *Safety Rules for Painting and Finishing* for additional emphasis.
2. Explain that paints, stains, and varnishes contain ingredients that could cause serious health problems if splashed into unprotected eyes or onto exposed skin.
3. Caution students about the health hazards connected with the inhaling of toxic fumes. Stress the need for adequate ventilation when working in closed areas.
4. Call special attention to the danger of smoking in situations where flammable liquids are present or where a combination of smoke and airborne dust might cause combustion.
5. When preparing older structures for repainting, review the procedures for testing for the presence of lead in old paint.

PAINT BRUSHES

1. Using Reproducible Master 20-2, *Parts of a Paint Brush*, identify the parts of a paint brush. Mention that quality brushes have flagged bristles so that

they can hold more paint and provide better coverage.

2. Discuss both natural bristles and synthetic bristles. Inform the students that synthetics are best for water-based paints and finishes.

3. Discuss the various types of brushes and the special uses of each. During this presentation, use Reproducible Master 20-3, *Types of Brushes*, to illustrate the various types. If possible, have several types of brushes available to pass among the students.

4. Display and demonstrate the use of wire brushes.

PAINT PANS, ROLLERS, AND PADS

1. Refer to Figures 20-5 and 20-6 as you introduce these painting tools. Explain why roller covers are made in both long and short naps. Also discuss the wide range of uses for rollers.

2. Compare paint pads to brushes and rollers. Explain the advantages and disadvantages and discuss the uses of paint pads.

3. Demonstrate the proper use of the paint brush (how to hold it for painting various surfaces, etc.)

4. Demonstrate the proper use of the pan to hold paint and distribute the paint evenly on the roller.

5. Demonstrate the proper use of the pan and paint pad.

MECHANICAL SPRAYING EQUIPMENT

1. Using Reproducible Master 20-4, *Parts of a Spray Gun*, name the various parts of a spray gun and explain the principle of operation.

2. Note that there are two different types of spray guns: suction type and pressure type. Explain how each works.

3. Have students examine parts of a spray gun as you disassemble it. Note the difference between an internal mix and an external mix spraying head (referring also to Figure 20-10).

4. Have students observe proper cleaning of the spray gun. Stress the importance of this step in maintaining the unit in proper working condition.

LADDERS AND SCAFFOLDS

1. Refer to Unit 29 while introducing students to safe use of scaffolds and ladders.

ELECTRIC SANDERS

1. Refer to Unit 4 as you review with students the types of electric sanders and the special uses of each.

2. Demonstrate proper use of each type.

3. Demonstrate proper technique for installing sandpaper on the sander.

HAND SANDERS

1. Discuss the various types of abrasives available and show samples of sand paper, explaining the uses of each.

2. Demonstrate procedure for attaching sandpaper to various types of hand sanders used in your program.

PAINTS, VARNISHES, AND STAINS

1. Define and compare these coatings and explain their differences.

2. Define the terms "paint" and "coatings," and explain their differences.

3. Name and explain the role of each ingredient in paint.

4. Introduce clear coatings and their relationship to stains.

5. Discuss fillers and their uses.

6. Explain the purpose of stains and demonstrate their effect on various types of woods.

7. Compare the three types of stains: spirit stains, oil stains, and water-based stains.

COLOR SELECTION

1. Define the nine color selection terms and their relationship to the color wheel.

2. Explain that paints may be selected ready-mixed from the factory or mixed to the customer's specifications at the paint store.

PREPARING SURFACES FOR COATINGS

1. Stress the need to properly prepare surfaces before applying a finish.

2. Discuss various surface conditions that must be corrected before finishes can be applied.

INSIDE PAINTING

1. Explain the importance of protecting surfaces that are *not* being painted. List the various ways of providing protection.

2. Detail steps that must be taken to prepare surfaces (such as removing grease and soil, deglossing surfaces, etc.). Note that deglossing can be done with chemical preparations, as well.

3. Review with students the proper order for painting an entire room; for painting trim; for painting a door; for painting a window.

EXTERIOR PAINTING

1. Have students prepare new wood for painting. Impress upon them the need to properly prepare

the new surfaces. Knots and pitch-containing areas are to be covered with a shellac or knot sealer.

2. Discuss problems that may be encountered when working with previously painted surfaces. Refer to Figure 20-33 as you discuss causes of paint failure.

3. If "live work" is available, have students prepare previously painted surfaces using tools available in your shop. Demonstrate the use of various tools in removing deteriorated surfaces.

4. Demonstrate proper use of brushes, rollers, and pads in applying paint. Discuss the differences in application methods between oil paints and water-based paints.

5. If spray equipment is available, demonstrate the proper procedures for preparing paint and for spraying.

WORKING WITH STAINS AND CLEAR FINISHES

1. Discuss surface preparation including lifting of dents, filling of holes, repair of minor defects, and sanding. Stress the importance of sanding to the appearance of the finished surface.

2. Explain the purpose and practice of bleaching.

3. Demonstrate the application of stains and clear finishes. Have students apply finishes to prepared surfaces.

ESTIMATING

1. Have students measure a shop wall and compute the area.

2. Next, have them read the label on a paint container to find the coverage of the paint.

3. Have students estimate the amount of paint needed to coat the wall.

PROBLEMS WITH COATINGS

1. Collect photos or samples of defects in old coatings. Have students determine the type of failure and its cause and how it could have been avoided.

2. In a similar manner, collect examples of problems in clear finishes. Have students study them and suggest their causes.

HANGING WALL COVERINGS

1. Review the types of wall paper used today and list their important characteristics.

2. Explain that wall paper is measured by the roll with a single roll varying from 27 to 29 sq. ft.

3. Have students take room measurements and, using the chart in Figure 20-34, determine the number of rolls needed for coverage.

4. Discuss the tools and other materials needed for wall papering and explain their use.

5. Describe the wall preparations needed before hanging paper. Include the following areas in your lecture.
   A. Stripping of old coverings.
   B. Use of sizing on new plaster or drywall.
   C. Drawing a plumb line as a guide to hanging the first strip of wall paper.

6. Describe the procedure of preparing a strip of wallpaper for hanging.

7. Explain the term, "booking," and demonstrate the technique.

8. Demonstrate the technique for hanging the strips.

UNIT REVIEW

1. Review the unit objectives. Be sure that students understand each objective.

2. Assign *Important Terms*, *Test Your Knowledge* questions and, *Outside Assignments* on pages 576-577 of the text. Review the answers in class.

3. Assign pages 145-148 of the workbook. Review the answers in class.

EVALUATION

1. Use Unit 20 Quiz for in-class evaluation. Correct the quizzes and return them to the students for review.

ANSWERS TO TEST YOUR KNOWLEDGE TEXT PAGE 576

1. D. All of the above.
2. False.
3. hog bristles
4. Wire
5. fast-drying
6. External mix.
7. Pigments
8. Filler.
9. primary
10. less
11. To protect areas that are not being painted.
12. First the ceiling, then walls, trim, and doors.
13. cutting in
14. B. seal knots and pitch pockets
15. Area to be covered and coverage of the coating.
16. elasticity
17. Cotton, linen, hemp, wood, and wastepaper.

ANSWERS TO WORKBOOK QUESTIONS
PAGES 145-148 ▬▬▬▬▬▬▬

1. A. Varnish and enamel.
   B. Flat trim.
   C. Oval sash.
   D. Angular sash.
   E. Flat wall.
2. A. Bristles or stock.
   B. Ferrule
   C. Handle
   D. Setting.
   E. Plug.
3. B. create a void that will hold a supply of paint
4. suction
5. A. Orbital
   B. vibrating
6. A. alcohol
   B. acetone
7. B. To provide a slightly roughened surface for better adhesion of the next coat.
8. Value
9. B. provide a place to strike off excess paint from a brush or roller
10. True.
11. False.
12. A. water
13. B. water under coating is pushing paint film away from wood surface
14. 1.8 or 2 gal.

ANSWERS TO UNIT 20 QUIZ ▬▬▬▬▬▬

1. False.
2. False.
3. True.
4. True.
5. internal mix
6. Vehicles
7. Fillers
8. primary
9. primer, sealer
10. bleaching
11. holiday
12. C
13. E
14. B
15. D
16. A

ANSWERS TO SECTION 4 EXAM ▬▬▬▬▬

1. True.
2. False.
3. False.
4. True.
5. False.
6. True.
7. False.
8. True.
9. True.
10. True.
11. False.
12. True.
13. False.
14. C. degree day
15. B. dew point
16. D. Polystyrene
17. C. decibel
18. A. carpenter
19. B. kicker
20. C
21. A
22. B
23. D
24. E
25. C
26. B
27. A
28. D
29. E
30. C
31. B
32. A
33. E
34. B
35. C
36. A
37. F
38. D
39. D
40. F
41. A
42. B
43. E
44. C
45. impact sounds
46. Sound
47. clear
48. balustrade
49. hand
50. baseboard
51. contact cement
52. Sound absorption
53. Backing board
54. plywood
55. 17-18
56. starting newel

Reproducible Master 20-1
Safety Rules for Painting and Finishing

Name _____ Period _____

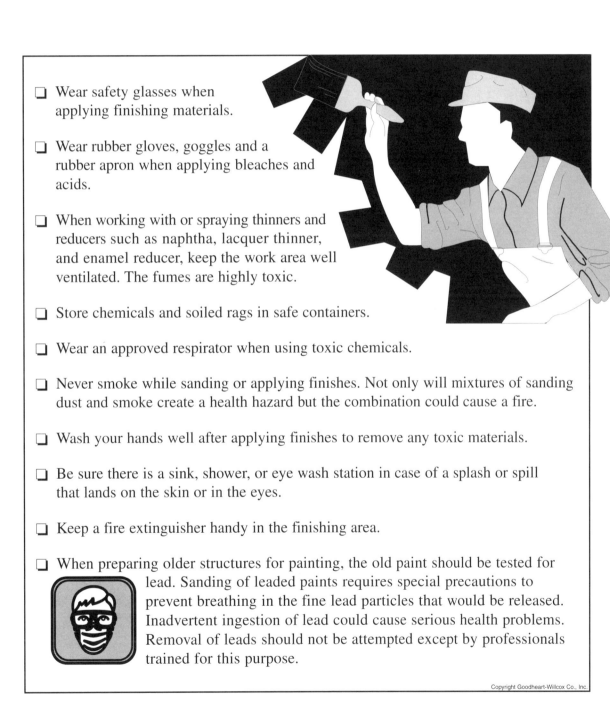

❑ Wear safety glasses when applying finishing materials.

❑ Wear rubber gloves, goggles and a rubber apron when applying bleaches and acids.

❑ When working with or spraying thinners and reducers such as naphtha, lacquer thinner, and enamel reducer, keep the work area well ventilated. The fumes are highly toxic.

❑ Store chemicals and soiled rags in safe containers.

❑ Wear an approved respirator when using toxic chemicals.

❑ Never smoke while sanding or applying finishes. Not only will mixtures of sanding dust and smoke create a health hazard but the combination could cause a fire.

❑ Wash your hands well after applying finishes to remove any toxic materials.

❑ Be sure there is a sink, shower, or eye wash station in case of a splash or spill that lands on the skin or in the eyes.

❑ Keep a fire extinguisher handy in the finishing area.

❑ When preparing older structures for painting, the old paint should be tested for lead. Sanding of leaded paints requires special precautions to prevent breathing in the fine lead particles that would be released. Inadvertent ingestion of lead could cause serious health problems. Removal of leads should not be attempted except by professionals trained for this purpose.

Parts of a Paint Brush

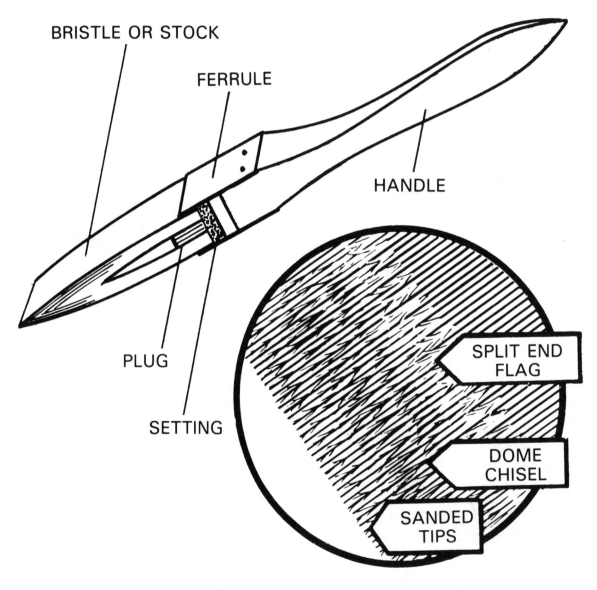

BRISTLE OR STOCK

FERRULE

HANDLE

PLUG

SETTING

SPLIT END FLAG

DOME CHISEL

SANDED TIPS

Reproducible Master 20-3
Types of Brushes

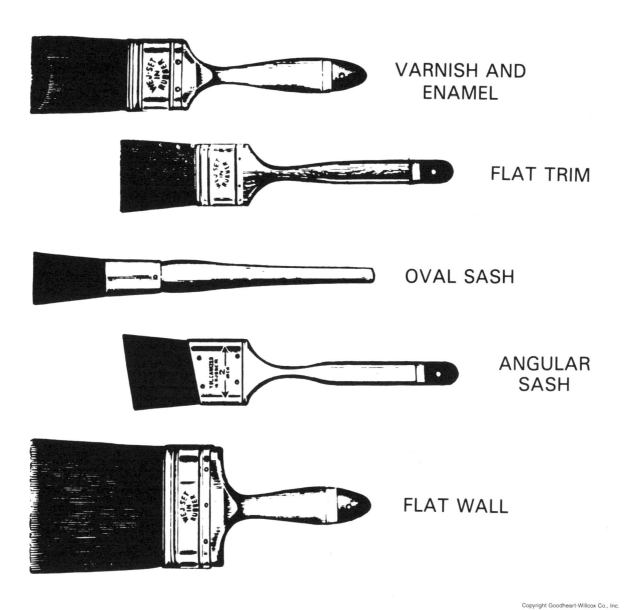

VARNISH AND
ENAMEL

FLAT TRIM

OVAL SASH

ANGULAR
SASH

FLAT WALL

Reproducible Master 20-4
Parts of a Spray Gun

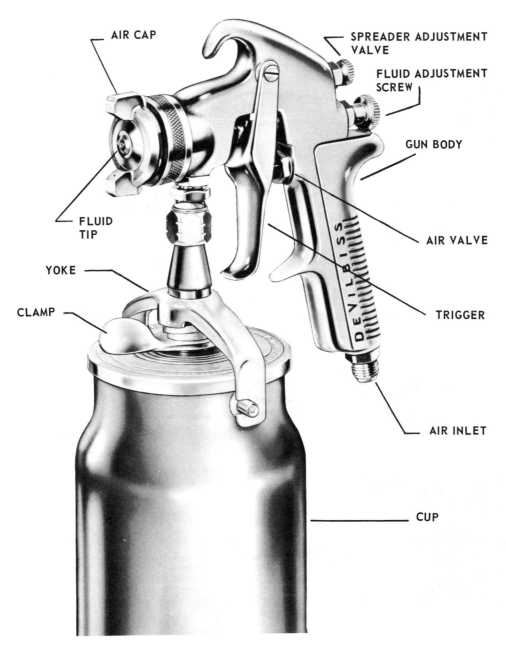

AIR CAP

SPREADER ADJUSTMENT VALVE

FLUID ADJUSTMENT SCREW

GUN BODY

FLUID TIP

AIR VALVE

YOKE

CLAMP

TRIGGER

AIR INLET

CUP

Unit 20 Quiz

Painting, Finishing, and Decorating

Name _____ Score _____

True-False

Circle T if the answer is True or F if the answer is False.

T F 1. Coatings is a term used to cover all types of finishes, except anodizing.

T F 2. It is safe for anyone to remove paint containing lead as long as a respirator is worn.

T F 3. Quality paint brushes taper from the ferrule to the tip of the stock.

T F 4. A suction-feed spray gun works best with light-bodied coating materials.

Completion

Fill in the correct word(s) that best completes the sentence.

_____ 5. A(n) _____ _____ spray gun works best when air pressures are low.

_____ 6. _____ are the oils or resins that make a paint fluid.

_____ 7. _____ are heavy-bodied liquids that fill the depressions of an open grain wood.

_____ 8. Red, blue, and yellow are known as the _____ colors.

_____ 9. On new exterior wood the first coat of finish should be a(n) _____ or a(n) _____ .

_____ 10. Wood _____ is the removal of some of a wood's color.

_____ 11. A(n) _____ is a bare spot in a clear finish.

Identification

_____ 12. Handle.

_____ 13. Plug.

_____ 14. Ferrule.

_____ 15. Setting.

_____ 16. Bristle or stock.

Section 4 Exam
Finishing

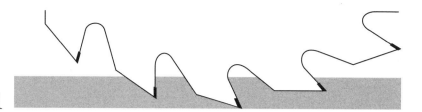

Name _____ Score _____

True-False

Circle T if the answer is True or F if the answer is False.

T F 1. Insulation with a higher R-value provides better insulating qualities than insulation with a lower value.

T F 2. A vapor barrier should be applied to the cold side of a wall.

T F 3. A portable circular saw must be used to cut drywall.

T F 4. Veneer plaster should be applied in thin layers, less than 1/8″ thick.

T F 5. In three-coat plaster work, the first coat is called the brown coat.

T F 6. When installing wood strip flooring, nails for the finish flooring should go through the subfloor and into the joists when possible.

T F 7. Underlayment is not required for flooring materials such as vinyl and linoleum.

T F 8. In stair construction, the number of treads is always one less than the number of risers.

T F 9. The doorframe forms the lining of the door opening, and also covers the edges of the partition.

T F 10. Gains are recesses used for hinges.

T F 11. The standard height for base kitchen cabinets is 30″.

T F 12. The facing is finished strips applied to the front of the cabinet frame.

T F 13. Flush drawers have a rabbet along the top and sides of the front.

Multiple Choice

Choose the answer that correctly completes the statement. Write the corresponding letter in the space provided.

_____ 14. A(n) _____ is the product of one day and the number of degrees Fahrenheit the mean temperature is below 65°F.
  A. R-value
  B. Btu
  C. degree day
  D. U-value

_____ 15. The temperature at which moisture is released as condensation when warm, moist air is cooled is called the _____.
  A. degree day
  B. dew point
  C. R-value
  D. None of the above.

_____ 16. _____ insulation is a rigid-type insulation that is commonly used for the perimeter of structures.

    A. Polyethylene
    B. Polyurethane
    C. Polyester
    D. Polystyrene

_____ 17. The unit of measure used to indicate the intensity or loudness of sound is called a(n) _____.

    A. impact sound
    B. masking sound
    C. decibel
    D. STC

_____ 18. When cabinets are constructed on the job, the _____ is responsible for determining the kind and size of joints.

    A. carpenter
    B. architect
    C. owner
    D. contractor

_____ 19. In drawer construction, a _____ is used to prevent the drawer from tilting forward when it is opened.

    A. leveler
    B. kicker
    C. baluster
    D. guide

Matching

Select the correct answer from the list on the right and place the corresponding letter in the blank on the left.

_____ 20. Conduction.

_____ 21. Convection.

_____ 22. Radiation.

A. Transfer of heat by another agent, usually air.
B. Transfer of heat through wave motion.
C. Transfer of heat from one molecule to another within a material, or from one material to another when they are held in direct contact with one another.

_____ 23. Btu.

_____ 24. k.

_____ 25. C.

_____ 26. R.

_____ 27. U.

A. Represents the total heat transmission.
B. Reciprocal of conductivity.
C. Conductance of a material.
D. Amount of heat needed to raise one pound of water one degree Fahrenheit.
E. Amount of heat transferred in one hour through one square foot of a given material that is 1″ thick and has a temperature difference between its surfaces of one degree Fahrenheit.

Name _____

_____ 28. Unit rise.

_____ 29. Unit run.

_____ 30. Headroom.

_____ 31. Nosing.

_____ 32. Stringer.

A. Support for a stairway.
B. Part of a stair that extends beyond the riser.
C. Vertical space above the stairway.
D. Height of a riser.
E. Width of a tread.

Identification

Identify the parts of the panel door.

_____ 33. Stile.

_____ 34. Mullion.

_____ 35. Panel.

_____ 36. Top rail.

_____ 37. Bottom rail.

_____ 38. Lock rail.

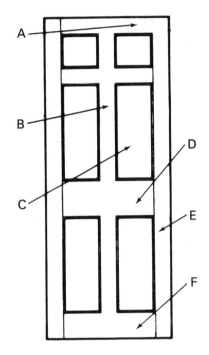

Identify the parts of the stairway.

_____ 39. Starting step.

_____ 40. Baluster.

_____ 41. Handrail.

_____ 42. Newel.

_____ 43. Open stringer.

_____ 44. Closed stringer.

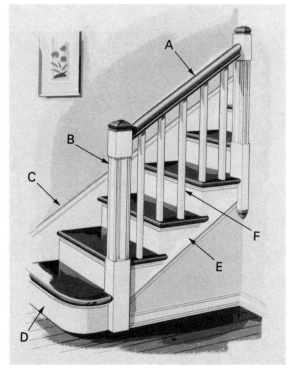

Completion

Place the answer that correctly completes the statement in the space provided.

_____ 45. Sounds that are carried through a building by the vibrations of the materials themselves are called _____.

_____ 46. _____ is a vibration or wave that can be heard.

_____ 47. The best grade of plain-sawed oak flooring is _____.

_____ 48. A(n) _____ is a decorative structure that supports the handrail for open main stairways.

_____ 49. The swing of a door is also called the _____ of the door.

_____ 50. A(n) _____ covers the joints between the wall surface and the floor.

_____ 51. Plastic laminates are commonly adhered to a base material with _____ _____.

_____ 52. _____ _____ is the capacity of a material or object to reduce the sound waves by absorbing them.

_____ 53. _____ _____ is similar to regular gypsum board, except that the covering is a gray liner paper.

_____ 54. Most carpenters prefer to use _____ as a flooring underlayment.

_____ 55. The sum of one riser and one tread should equal _____ inches.

_____ 56. When constructing open stairways, the _____ _____ must be securely anchored to the starter step or carried down through the floor and attached to the building frame.

21

Chimneys and Fireplaces

OBJECTIVES

Students will be able to:

❏ Explain how masonry chimneys are constructed around flue linings.

❏ Name the parts of a typical masonry fireplace.

❏ Describe procedures for the construction of the chimney, hearth, walls, and throat.

❏ Define function of damper and smoke shelf.

❏ Calculate the required flue area for a given fire place.

❏ Describe common types of factory-built fire places.

❏ List special considerations for installing factory-built fireplace units.

❏ Install a prefabricated flue.

INSTRUCTIONAL MATERIALS

Text: Pages 581-594
Important Terms, page 593
Test Your Knowledge, page 593
Outside Assignments, page 594

Workbook: Pages 149-154

Instructor's Manual:
Reproducible Master 21-1, *Masonry Fireplace Parts*
Unit 21 Quiz

TRADE-RELATED MATH

Ratios can be used to compare relationships between numbers. For example, one method of sizing flues for fireplaces is to allow 13 square inches of flue area for every 1 square foot of fireplace opening. If you want to determine the flue size that is appropriate for a fireplace with a 9 square foot opening, the ratio can be set up as follows:

$$\frac{13 \text{ sq. in.}}{1 \text{ sq. ft.}} = \frac{x \text{ sq. in.}}{9 \text{ sq. ft.}}$$

$$(13 \times 9) \div 1 = 117 \div 1 = 117 \text{ sq. in.}$$

INSTRUCTIONAL CONCEPTS AND STUDENT LEARNING EXPERIENCES

MASONRY CHIMNEYS

1. Explain that masonry chimneys neither receive support from nor give support to the building frame.

2. Discuss the factors that determine the size of a chimney.

3. Cite reasons for extending chimneys above the roof line of a structure. Refer to Figure 21-1 for examples of the minimum requirements for chimney heights above roof lines. Have the students refer to the local building code to determine if these heights are similar to your locality.

4. Explain that combustible materials such as wood framing should be located at least 2″ away from the chimney wall. List the types of materials that can be used to fill this open space.

5. Discuss the purpose of the flue lining.

6. Have the students identify the basic shapes of flue linings. Explain that there are different methods of measurement for the three types.

7. Describe the procedure typically used for constructing a masonry chimney. Mention that the chimney wall is usually constructed around the flue lining.

8. Discuss the reasons for corbeling a chimney.

9. Review the type of flashing that is commonly used around a chimney. Mention that one part of the flashing is built into the chimney while the other part is attached to the surface of the roof.

MASONRY FIREPLACES

1. Using Reproducible Master 21-1, Masonry Fireplace Parts, identify the primary parts. Describe the purpose of each part as it is identified.

2. Refer to Figure 21-6 or a stock plan and point out

the fireplace details that are important for a carpenter to notice.

3. Discuss the criteria that can be used to determine the appropriate size of fireplace for a given structure. Refer to Figure 21-7 for dimensions that are commonly used for fireplaces.
4. Describe the construction of a hearth. Discuss variations of the basic design for slab-on-grade structures.
5. Illustrate the design of the side and back walls. Mention that the total thickness of the walls, including the firebrick, should not be less than 8″.
6. Define "splay" and describe its purpose.
7. Identify the two parts of the damper. Discuss how the damper affects the downdraft.
8. Review the purposes of the smoke shelf and smoke chamber. Describe possible scenarios that may occur if a gust of wind passes over the flue, or if the smoke shelf and chamber were omitted from a design.
9. Discuss the methods used to determine the cross-sectional area of a flue. Mention that for chimneys over 20′ tall that the flue size should be increased.
10. Discuss a typical construction sequence for masonry fireplaces.
11. Using photographs from architectural magazines, show the students modern fireplace designs. Discuss how the flue area is determined for these types of fireplaces.
12. Discuss the advantages of built-in circulators. Describe the construction of a fireplace around a circulator.

PREFABRICATED CHIMNEYS

1. Discuss the construction of prefabricated chimneys. Explain the function of a carpenter when installing these types of chimneys.
2. Stress the importance of installing prefabricated chimneys that are approved by the Underwriters' Laboratories (UL) or other recognized associations.

PREFABRICATED FIREPLACES

1. Compare the operation of a prefabricated fireplace to that of a built-in circulator.
2. Define the term "zero-clearance" and how it applies to prefabricated fireplaces.
3. Describe the frame construction around a prefabricated fireplace. Stress that the manufacturer's recommendations should be followed when constructing the framing.
4. Refer to Figure 21-27 to illustrate the type of

chimney system commonly used for prefabricated fireplaces.
5. Define "chase" and describe how it can be framed and constructed.

GLASS ENCLOSURES

1. Explain the purpose of glass enclosures for fireplaces.

UNIT REVIEW

1. Review the unit objectives. Be sure that the students fully understand each objective.
2. Assign *Important Terms, Test Your Knowledge* questions, and *Outside Assignments* on pages 593-594 of the text. Review the answers in class.
3. Assign pages 149-154 of the workbook. Review the answers in class.

EVALUATION

1. Use Unit 21 Quiz for in-class evaluation. Correct the quizzes and return them to the students for review.

ANSWERS TO TEST YOUR KNOWLEDGE TEXT PAGE 593

1. flues
2. 2″
3. 60
4. damper
5. width
6. cement/mortar
7. 1/10
8. triple
9. stainless steel
10. burning
11. floor

ANSWERS TO WORKBOOK QUESTIONS PAGES 149-154

1. A. 6″
   B. 4″
2. A. 2′-6″
   B. 2′-0″
   C. 3′-0″
3. A. 2″
   B. 7″
4. B. 5/8″
5. B. 12″ dia.
6. D. 60°
7. C. 6″

8. A. Flue.
    B. Damper.
    C. Smoke shelf.
    D. Firebrick.
    E. Hearth.
    F. Lintel.
    G. Smoke chamber.
9. ash pit
10. B. lintel
11. C. 28″ to 30″
12. A. damper
    B. 8″
13. C. 5″ per foot
14. B. 6″ to 8″
15. D. expansion
16. B. 1/2″
17. D. 13 sq. in.
18. larger
19. A. 4″
20. smoke chamber
21. A. 3 1/2″

22. A. Warm air is fed into the room from grillwork along the bottom edge.
23. A. single-walled
24. chase

ANSWERS TO UNIT 21 QUIZ

1. B. damper
2. D. 2′-0″
3. A. 4″
4. C. smoke chamber
5. B. 3 1/2″
6. J
7. I
8. F
9. G
10. A
11. D
12. C
13. H
14. B
15. E

Reproducible Master 21-1
Masonry Fireplace Parts

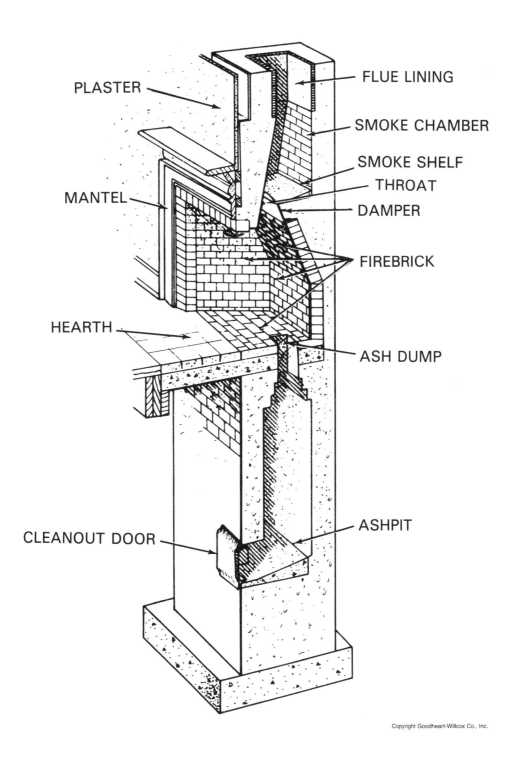

PLASTER

FLUE LINING

SMOKE CHAMBER

SMOKE SHELF

THROAT

MANTEL

DAMPER

FIREBRICK

HEARTH

ASH DUMP

CLEANOUT DOOR

ASHPIT

Unit 21 Quiz
Chimneys and Fireplaces

Name _____ Score _____

Multiple Choice

Choose the answer that correctly completes the statement. Write the corresponding letter in the space provided.

_____ 1. The _____ is used to control a fire, and also prevent loss of heat from a room when the fireplace is not being operated.

    A. smoke shelf
    B. damper
    C. throat
    D. cleanout door

_____ 2. A chimney should extend _____ above any roof ridge that is within a 10 foot horizontal distance.

    A. 6″
    B. 1′-0″
    C. 1′-6″
    D. 2′-0″

_____ 3. The flue lining should project at least _____ above the top brick course or cap.

    A. 4″
    B. 8″
    C. 1′-0″
    D. 1′-4″

_____ 4. The _____ is the space extending from the top of the throat to the bottom of the flue.

    A. flue lining
    B. hearth
    C. smoke chamber
    D. smoke shelf

_____ 5. According to FHA specifications, wooden parts of a prefabricated chimney should not be placed closer than _____ from the edge of the opening.

    A. 1 1/2″
    B. 3 1/2″
    C. 5 1/2″
    D. 7 1/2″

Identification

Identify the parts of the masonry fireplace.

_____   6.   Mantel.

_____   7.   Hearth.

_____   8.   Ash dump.

_____   9.   Ashpit.

_____   10.   Flue lining.

_____   11.   Throat.

_____   12.   Smoke shelf.

_____   13.   Cleanout door.

_____   14.   Smoke chamber.

_____   15.   Damper.

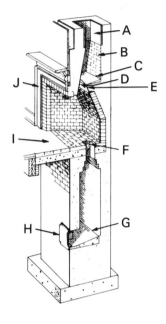

22

Post-and-Beam Construction

OBJECTIVES

Students will be able to:

❏ List the advantages and disadvantages of post-and-beam construction.

❏ Describe general specifications for supporting posts.

❏ Compare transverse and longitudinal beams.

❏ Describe how roof and floor planks should be elected and installed.

❏ Sketch basic construction details of stressed skin panels and box beams.

INSTRUCTIONAL MATERIALS

Text: Pages 595-610

Important Terms, page 610

Test Your Knowledge, page 610

Outside Assignments, page 610

Workbook: Pages 155-160

Instructor's Manual:

Reproducible Master 22-1, *Post-and-Beam Framing Details, Parts A* and *B*

Reproducible Master 22-2, *Plank-and-Beam Roof Construction*

Unit 22 Quiz

TRADE-RELATED MATH

In many cases, a carpenter is required to read tables and charts. Span tables can be used to determine the size of beams required for a given span. For example, with the table below, it can be determined that a W8 × 10 steel beam is required to support a 10,400 lb. load when a 12′ span is required.

INSTRUCTIONAL CONCEPTS AND STUDENT LEARNING EXPERIENCES

ADVANTAGES

1. Compare typical post-and-beam construction with conventional framing. Have students list the advantages of post-and-beam construction. Also cite limitations of this construction method.

FOUNDATIONS AND POSTS

1. Discuss the two basic types of foundations that

DESIGNATION WT./FT.	NOMINAL SIZE DP. x WD.	SPAN IN FEET									
		8′	10′	12′	14′	16′	18′	20′	22′	24′	26′
W8x10	8x4	15.6	12.5	10.4	8.9	7.8	6.9	—	—	—	—
W8x13	8x4	19.9	15.9	13.3	11.4	9.9	8.8	—	—	—	—
W8x15	8x4	23.6	18.9	15.8	13.5	11.8	10.5	—	—	—	—
W8x18	8x5 1/4	30.4	24.3	20.3	17.4	15.2	13.5	—	—	—	—
W8x21	8x5 1/4	36.4	29.1	24.3	20.8	18.2	16.2	—	—	—	—
W8x24	8x6 1/2	41.8	33.4	27.8	23.9	20.9	18.6	—	—	—	—
W8x28	8x6 1/2	48.6	38.9	32.4	27.8	24.3	21.6	—	—	—	—
W10x22	10x5 3/4	—	—	30.9	26.5	23.2	20.6	18.6	16.9	—	—
W10x26	10x5 3/4	—	—	37.2	31.9	27.9	24.8	22.3	20.3	—	—
W10x30	10x5 3/4	—	—	43.2	37.0	32.4	28.8	25.9	23.6	—	—
W12x26	12x6 1/2	—	—	—	—	33.4	29.7	26.7	24.3	22.3	20.5
W12x30	12x6 1/2	—	—	—	—	38.6	34.3	30.9	28.1	25.8	23.8
W12x35	12x6 1/2	—	—	—	—	45.6	40.6	36.5	33.2	30.4	28.1

can be used for post-and-beam construction—continuous walls or piers under each post.

2. Identify the minimum size posts that are recommended. Emphasize that these posts must not only be able to support the load, but also provide bearing surfaces for the ends of the beams. Mention that as the post height increases, the cross-sectional area must also increase.

3. Using part *A* of Reproducible Master 22-1, *Post-and-Beam Framing Details*, discuss the purpose of bearing blocks in post-and-beam construction.

4. Using part *B* of reproducible Master 22-1, *Post-and-Beam Framing Details*, compare the use of plates in post-and-beam framing construction to that of plates in conventional framing.

FLOOR BEAMS

1. Describe the type of floor beams that can be used for post-and-beam construction.

2. Explain how the silhouette of the structure can be kept low by forming beam pockets in the foundation walls.

BEAM DESCRIPTIONS

1. Discuss techniques that can be used to improve the appearance of exposed beams.

ROOF BEAMS

1. Using Reproducible Master 22-2, *Plank-and-Beam Roof Construction,* identify the two types of beams used to support roof systems. Discuss the purpose of each type.

2. Describe the techniques used to secure transverse and ridge beams into position.

FASTENERS

1. Stress that conventional nailing patterns *cannot* be used for post-and-beam construction. Identify the types of metal fasteners that can be used.

PARTITIONS

1. Discuss the problems that may be encountered when trying to erect partitions in post-and-beam construction. Mention that it is best to erect these partitions after the beams and planks are in position.

2. Describe methods for installing partitions that run parallel to transverse beams.

PLANKS

1. Identify the plank patterns that are commonly used. Note that when end-matched planks are used, they need not meet over beams.

2. Stress the importance of selecting materials (floor planks, roof planks, beams, and posts) that will correspond closely to the moisture content of the interior of the structure. Have the students identify problems that may occur if the moisture content varies a great deal.

3. Describe how insulation and a vapor barrier is installed for plank roof structures.

STRESSED SKIN PANELS

1. Explain the construction of stressed skin panels. Discuss advantages of using these panels.

BOX BEAMS

1. Discuss advantages of box beams in residential construction.

2. Describe the construction of box beams.

LAMINATED BEAMS AND ARCHES

1. Identify the basic types of laminated wood arches shown in Figure 22-31. Have the students note the support and lateral thrust that must be provided.

UNIT REVIEW

1. Review the unit objectives. Be sure that the students fully understand each objective.

2. Assign *Important Terms, Test Your Knowledge* questions, and *Outside Assignments* on page 610 of the text. Review answers in class.

3. Assign pages 155-160 of the workbook, again reviewing answers in class.

EVALUATION

1. Use Unit 22 Quiz for in-class evaluation. Correct the quizzes and return them to the students for review.

ANSWERS TO TEST YOUR KNOWLEDGE
TEXT PAGE 610

1. smallest
2. transverse
3. 40
4. 4
5. sandwich
6. webs
7. False.

ANSWERS TO WORKBOOK QUESTIONS
PAGES 155-160

1. A. Planks.
   B. Beam.
   C. Plate.
   D. Beam.
   E. Post.
2. fire resistance
3. C. bearing blocks
4. 34
5. 30
6. A. Solid.
   B. Vertical laminated.
   C. Horizontal laminated.
   D. Spaced with wood blocking.
7. A. 16″
8. A. Sole plate.
   B. Header.
   C. Plank.
   D. Beam.
   E. Sill.
9. longitudinal
10. lag screws
11. D. sloping
12. is
13. C. 2″ to 4″
14. A. Vapor barrier.
    B. Rigid insulation.
    C. Built-up roof.
    D. Gravel stop.
15. stressed skin
16. sandwich
17. D. 120′
18. softwood
19. A. Three-centered.
    B. A-frame.
    C. Tudor.
    D. Parabolic.
    E. Radial.
    F. Gothic.
20. B. 24″

ANSWERS TO UNIT 22 QUIZ

1. B. 4×4
2. D. All of the above.
3. D. Both B and C.
4. C. Stressed skin
5. D
6. E
7. F
8. A
9. B
10. C
11. G
12. D
13. F
14. A
15. B
16. E
17. C

Modern Carpentry Instructor's Manual

Reproducible Master 22-1 (Part A)
Post-and-Beam Framing Details

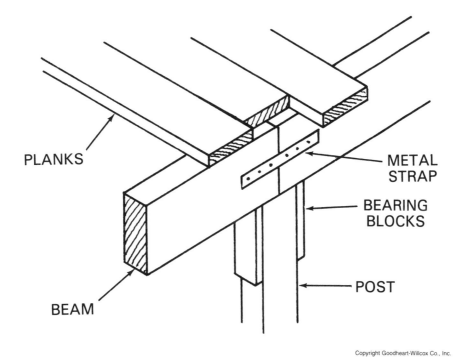

PLANKS

METAL
STRAP

BEARING
BLOCKS

POST

BEAM

Reproducible Master 22-1 (Part B)
Post-and-Beam Framing Details

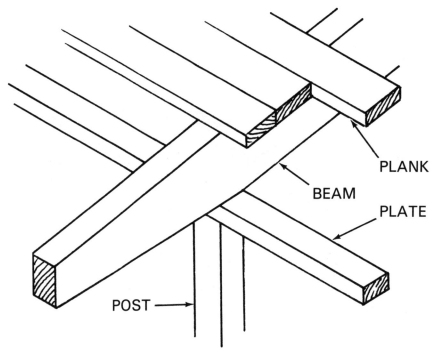

PLANK

BEAM

PLATE

POST

Reproducible Master 22-2

Plank-and-Beam Roof Construction

TRANSVERSE BEAM

LONGITUDINAL BEAM

Unit 22 Quiz

Post-and-Beam Construction

Name _____ Score _____

Multiple Choice

Choose the answer that correctly completes the statement. Write the corresponding letter in the space provided.

_____ 1. In general, posts should not have a nominal size less than _____.

       A. 2×4

       B. 4×4

       C. 8×8

       D. None of the above.

_____ 2. Beams for floor structures may be _____.

       A. solid

       B. glued-laminated

       C. built-up

       D. All of the above.

_____ 3. _____ beams run parallel to the supporting side walls and ridge beam.

       A. Transverse

       B. Longitudinal

       C. Purlin

       D. Both B and C.

_____ 4. _____ panels are made by gluing sheets of plywood to longitudinal frame members.

       A. Solid-core plywood

       B. Particle board

       C. Stressed skin

       D. All of the above.

Identification

Identify the parts of the typical sill construction for the post-and-beam frame.

_____ 5. Sill.

_____ 6. Foundation.

_____ 7. Header.

_____ 8. Post.

_____ 9. Plank.

_____ 10. Beam.

_____ 11. Sole plate.

Identify the types of laminated wood arches.

_____ 12. Tudor.

_____ 13. Parabolic.

_____ 14. Radial.

_____ 15. Gothic.

_____ 16. Three-centered.

_____ 17. A-frame.

23

Systems-Built Housing

OBJECTIVES

Students will be able to:
- ❏ Discuss the changes that have taken place in the technology of systems-built housing.
- ❏ Comment on the variety of factory-built components that are utilized in a systems-built home.
- ❏ List and differentiate between the basic types of systems-built buildings.
- ❏ Detail the erection sequence of a panelized home.
- ❏ Define terms used in the industry.
- ❏ List erection sequence for a paneled home.

INSTRUCTIONAL MATERIALS

Text: Pages 611-626
> *Important Terms,* page 625
> *Test Your Knowledge,* page 625
> *Outside Assignments,* page 625

Workbook: Pages 161-164

Instructor's Manual:
> Reproducible Master 23-1, *Panelized*
>   *Construction*
> Unit 23 Quiz

INSTRUCTIONAL CONCEPTS AND STUDENT LEARNING EXPERIENCES

INTRODUCTION
1. Define the new term, "systems-built."
2. Describe the technological changes that have occurred in the factory-built housing industry.

TRANSPORTING
1. Explain method of moving systems-built housing from factory to building site.
2. List types of buildings that are assembled or partially assembled in factory setup.

COMPONENTS
1. List the components commonly used in prefabricated construction. Cite reasons that these

components are factory built while others are not.
2. Explain the advantages gained through prefabrication of building components.

TYPES OF FACTORY-BUILT HOMES
1. Identify the five basic types of systems-built homes and briefly describe the construction of each.
2. Briefly discuss the four qualities offered by factory-building.
3. Review the factory cutting and assembly of a panelized home.
4. List the characteristics of a precut house.
5. List the characteristics of a panelized prefabrication, and explain how it differs from precut prefabrication.
6. List the characteristics of sectionalized (modular) prefabrication, and explain how it differs from panelized construction. Discuss the advantages that this type of construction offers over other prefabrication methods. Describe any limitations or disadvantages of sectionalized prefabrication.
7. List the characteristics of mobile home construction. Discuss advantages and disadvantages of this type of prefabrication.

ON-SITE CONSTRUCTION
1. Describe the general procedure for on-site construction of a modular prefabricated house.
2. Discuss disadvantages of prefabricated construction that occurs on the building site.

UNIT REVIEW
1. Review the unit objectives. Be sure that the students fully understand each objective.
2. Assign *Important Terms, Test Your Knowledge* questions, and *Outside Assignments* on page 625 of the text. Review the answers in class.
3. Assign pages 161-164 of the workbook. Review the answers in class.

EVALUATION

1. Use Unit 23 Quiz for in-class evaluation. Correct the quizzes and return them to the students for review.

ANSWERS TO TEST YOUR KNOWLEDGE
TEXT PAGE 625

1. Systems-built and factory-built.
2. Truss roof.
3. Flat stressed-skin panels and flat sandwich panels.
4. Modular (or mods)
5. closed
6. modular
7. precut
8. Foundation or slab has already been constructed and the systems-built house is on trailers.
9. False.
10. week
11. foundation
12. A. wet inside

ANSWERS TO WORKBOOK QUESTIONS
PAGES 161-164

1. building site
2. header

3. C. double-end
4. Modular: C
   Precut: B
   Manufactured home: D
   Log home: E
   Panelized: A
5. Girder truss/floor truss.
6. B. gang-nail
7. panelized
8. jigs
9. channels, chases
10. A. compressed air
11. C. Plumbing fixtures are seldom installed until erection is completed on the building site.
12. mechanical core
13. B. 4500 lb.
14. C. 2×3

ANSWERS TO UNIT 23 QUIZ

1. True.
2. False.
3. False.
4. F. All of the above.
5. C. modular
6. C. mechanical
7. A. closed panels

Reproducible Master 23-1

Panelized Construction

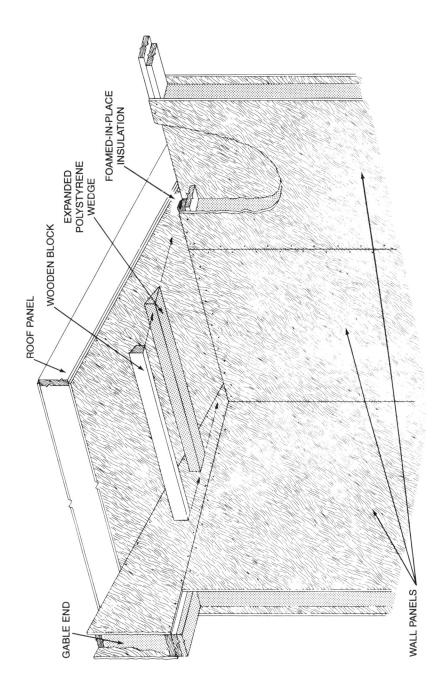

FOAMED-IN-PLACE INSULATION

EXPANDED POLYSTYRENE WEDGE

WOODEN BLOCK

ROOF PANEL

GABLE END

WALL PANELS

Unit 23 Quiz

Systems-Built Housing

Name _____ Score _____

True-False

Circle T if the answer is True or F if the answer is False.

T F 1. A systems-built house consists of components, panels, or modules cut and assembled in a factory.

T F 2. Firms producing systems-built structures do not offer custom designing.

T F 3. Modular units are easily lifted by hand onto their foundation.

Multiple Choice

Choose the answer that correctly completes the statement. Write the corresponding letter in the space provided.

_____ 4. A basic type of factory-built construction is the _____.

    A. precut
    B. panelized
    C. modular
    D. log home
    E. manufactured home
    F. All of the above.

_____ 5. In _____ systems construction, entire sections of the structure are built and finished in manufacturing plants.

    A. precut
    B. panelized
    C. modular
    D. log
    E. all of the above.

_____ 6. A modular section that has a group of plumbing and heating facilities is often called a(n) _____ core.

    A. electrical
    B. modular
    C. mechanical
    D. None of the above.

_____ 7. Prefabricated panels finished on both inside and outside are called _____ .

    A. closed panels
    B. open panels
    C. stressed skin panels
    D. mechanical cores

24

Passive Solar Construction

OBJECTIVES

Students will be able to:

❑ Define conduction, convection, radiation, and thermosiphoning.

❑ Explain the difference between passive and active solar construction.

❑ List and describe the three types of passive solar construction.

❑ Determine the amount of glazing and storage for a passive system.

❑ Locate a dwelling on a lot for maximum solar gain.

❑ Design and install various passive solar structures including a Trombe wall, wall and floor heat storage systems, protective overhangs, venting, and other fixed or movable controls.

INSTRUCTIONAL MATERIALS

Text: Pages 627-642

    Important Terms, page 640

    Test Your Knowledge, page 640

    Outside Assignments, page 641

Workbook: Pages 165-170

Instructor's Manual:

    Reproducible Master 24-1, *Trombe Wall Section*

    Reproducible Master 24-2, *Cross Section, Well-Insulated Dwelling*

    Unit 24 Quiz

TRADE-RELATED MATH

The width of the overhang is important when designing a solar structure. Three factors are used to determine the overhang distance: height of the window or collector, height difference between the header and the window or collector, and latitude of the construction site. For example, the window height might be 6'-8", the header height is 7'-6", and the construction

site is in southern Minnesota. The formula, used with the chart in Figure 24-17, is as follows:

$$\text{Projection} = \frac{\text{window height} + \text{height difference}}{\text{factor}}$$

$$\text{Projection} = \frac{6'\text{-}8'' + (7'\text{-}6'' - 6'\text{-}8'')}{2.1}$$

$$= \frac{6'\text{-}8'' + 10''}{2.1}$$

$$= \frac{7'\text{-} 6''}{2.1''}$$

$$= \frac{90}{2.1}$$

$$= 42.85$$

$$= 42\frac{27}{32}$$

$$\text{Projection} = 3'\text{-}6\,\frac{27''}{32}$$

INSTRUCTIONAL CONCEPTS AND STUDENT LEARNING EXPERIENCES

HOW RADIATION AND HEAT ACT

1. Define the term "greenhouse effect." Discuss its effect on solar construction.

2. Identify the three heat transfer methods—conduction, convection, and radiation.

3. Describe the concept of conduction as it relates to building materials.

4. Describe the concept of convection, stressing that this type of heat transfer only occurs in fluids or gases.

5. Describe the concept of radiation as it relates to building components.

6. Define "thermosiphoning" and how this concept is applied to both passive and active solar heating.

TYPES OF SOLAR CONSTRUCTION

1. Identify the two approaches used in solar construction—active and passive.

2. Describe an active solar system used in residential construction. Explain why it is called "active."
3. Using Figure 24-7, identify the typical components of an active solar energy system.

TYPES OF PASSIVE SOLAR ENERGY

1. Describe the difference between an active solar energy system and a passive solar energy system.
2. Define the term "gain" as it relates to solar construction.
3. Identify the three basic types of passive solar construction—direct gain, indirect gain, and isolated gain. Ask the students to list the architectural styles that work best with passive solar construction.
4. Discuss how a simple direct gain solar system works. (Be sure to include the terms "radiation" and "convection" in the discussion.) List disadvantages of this type of system.
5. Explain the difference between a direct gain and an indirect gain system.
6. Discuss the purpose of a Trombe wall, and describe its operation. Explain how the radiation and convection principles are used to distribute the heat.
7. Describe how a water storage wall works. Discuss the use of phase change materials instead of water.
8. Explain how an isolated gain system works. Have students list its advantages over other types of solar systems.

PASSIVE SOLAR ADVANTAGES

1. Discuss the advantages of passive solar systems over active solar systems.

PASSIVE SOLAR DISADVANTAGES

1. Discuss the disadvantages of passive solar systems.

SOLAR HEAT CONTROL

1. Identify methods used to control solar heat for passive solar systems. These might include: –structural features of the solar house, movable insulation, and natural shade obtained from trees and other surroundings.
2. List the three factors that are used to determine the width of the overhang for a solar structure.
3. Use the formula shown on page 633 to determine the overhang projection that is required in your area of the country.
4. Discuss the advantages of using movable insulation for solar structures. Identify various types of movable insulation that can be used and cite advantages or disadvantages of each type.
5. Explain the purpose of adequate venting in indirect and isolated gain systems.
6. Discuss the proper orientation of a solar structure for taking maximum advantage of the solar system.
7. Identify the three sources of energy that supply heat to a structure on a winter day. Discuss the appropriate location of appliances in a solar structure to balance internal heat gain.

BUILDING PASSIVE SOLAR STRUCTURES

1. Identify the materials commonly used to store heat in a solar structure. Cite advantages and disadvantages of each type.
2. Discuss how thermal storage systems are sized for solar structures. Explain how the thickness and color of the storage walls also affects the temperature inside a solar structure.
3. Using Reproducible Master 24-1, *Trombe Wall Section,* discuss the construction of a Trombe wall. Have the students note the use of the insulation in the hollow header and joist spaces. Explain the types of adjustments that must be made to the footing and foundation on which a Trombe wall is located.

SPECIAL CONCERNS

1. Mention that the thermal glazing should be constructed so that the glass panels can be removed for cleaning. Emphasize that double glazing is the most efficient type of glazing used in a solar structure.

DESIGNING THE ISOLATED GAIN SYSTEM

1. Discuss the design of an isolated gain system. Based on the sketch shown in Figure 24-13, have the students determine the most inexpensive, yet effective design.

PASSIVE THERMOSIPHON SYSTEM

1. Have the students determine the volume of the storage area required for a given size solar collector. Stress that the rock bins for thermal storage must be well insulated. Also explain that dampers should be used in the ductwork leading to the storage bin so that heat is not released back to the bin at night.

INSULATING PASSIVE SOLAR BUILDINGS

1. Using Reproducible Master 24-2, *Cross Section, Well-Insulated Dwelling,* emphasize the

importance of placing thermal barriers anywhere heat could leak out of the structure.

UNIT REVIEW

1. Review the unit objectives. Be sure that the students fully understand each objective.
2. Assign *Important Terms, Test Your Knowledge* questions, and *Outside Assignments* on pages 640-641 of the text. Review answers in class.
3. Assign pages 165–170 of the workbook. Review answers in class.

EVALUATION

1. Use Unit 24 Quiz for in-class evaluation. Correct the quizzes and return them to the students for review.

ANSWERS TO TEST YOUR KNOWLEDGE TEXT PAGE 640

1. greenhouse effect
2. Conduction, convection, and radiation.
3. Thermosiphoning
4. A. It has few, if any, moving parts that require mechanical or electrical energy for operation.
   B. It is usually a stationary, structural part of the house.
   D. Collection, storage, and transporting of heat is usually done naturally by the materials used in construction.
5. A. Control of the heat is not as responsive as either conventional heat systems or active solar systems.
   B. It is not a simple matter to control heat and heat distribution.
6. The height of the window or collector. The height of the header above the window or collector. The latitude of the construction site.
7. Orientation is how the house is placed in relation to the sun's position in the southern sky. Or, locating the building so one wall catches the sun's rays all day long.
8. appliances, occupants
9. 150 lb.
10. False.
11. B. 520 to 800 cu. ft.

ANSWERS TO WORKBOOK QUESTIONS PAGES 165-170

1. heat
2. A. Heat from solar radiation cannot pass through glass.
3. A. Overhang.
   B. Heat storage, wall.
   C. Heat storage, floor.
   D. Winter sun angle.
4. active
5. A. Thermal storage.
   B. Wall insulation.
   C. Glazing.
   D. Additional thermal storage.
   E. Vents.
6. C. System is expensive to maintain.
   D. System will not work during a power outage.
7. A. Trees/natural shade.
   B. Constructed protection.
   C. Movable insulation.
8. 5'-3"
9. C. 5.6
10. Movable insulation
11. E. locate the structure so the south wall will catch the sun's rays all day long
12. B. orientation
13. windows/glazing
14. north
15. C. 10"-14"
16. 813-1250 cu. ft.

ANSWERS TO UNIT 24 QUIZ

1. True.
2. False.
3. False.
4. True.
5. False.
6. False.
7. True.
8. D. All of the above.
9. B. indirect gain
10. B. south
11. C. isolated
12. D. All of the above.

Reproducible Master 24-1
Trombe Wall Section

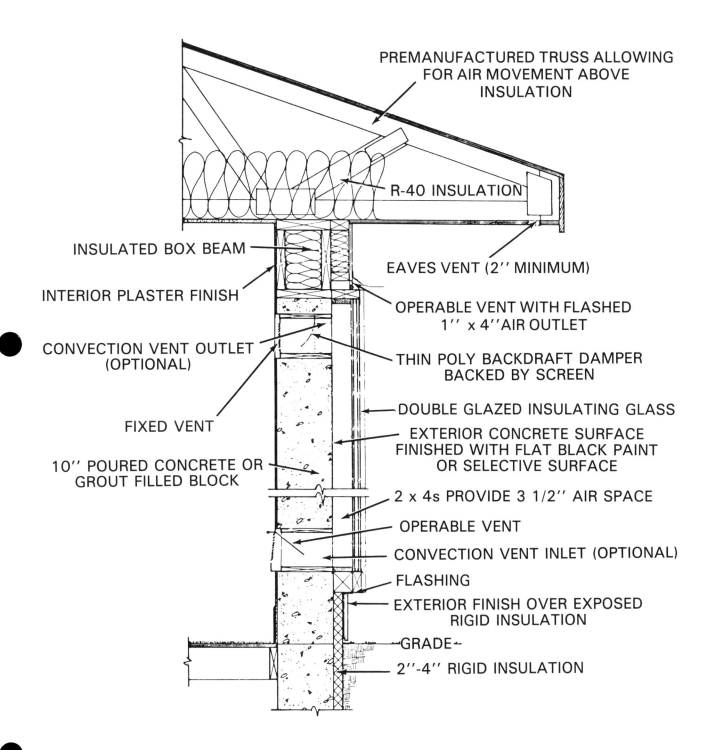

PREMANUFACTURED TRUSS ALLOWING FOR AIR MOVEMENT ABOVE INSULATION

R-40 INSULATION

INSULATED BOX BEAM

INTERIOR PLASTER FINISH

EAVES VENT (2'' MINIMUM)

OPERABLE VENT WITH FLASHED 1'' x 4'' AIR OUTLET

CONVECTION VENT OUTLET (OPTIONAL)

THIN POLY BACKDRAFT DAMPER BACKED BY SCREEN

FIXED VENT

DOUBLE GLAZED INSULATING GLASS

EXTERIOR CONCRETE SURFACE FINISHED WITH FLAT BLACK PAINT OR SELECTIVE SURFACE

10'' POURED CONCRETE OR GROUT FILLED BLOCK

2 x 4s PROVIDE 3 1/2'' AIR SPACE

OPERABLE VENT

CONVECTION VENT INLET (OPTIONAL)

FLASHING

EXTERIOR FINISH OVER EXPOSED RIGID INSULATION

GRADE

2''-4'' RIGID INSULATION

Reproducible Master 24-2
Cross Section, Well-Insulated Dwelling

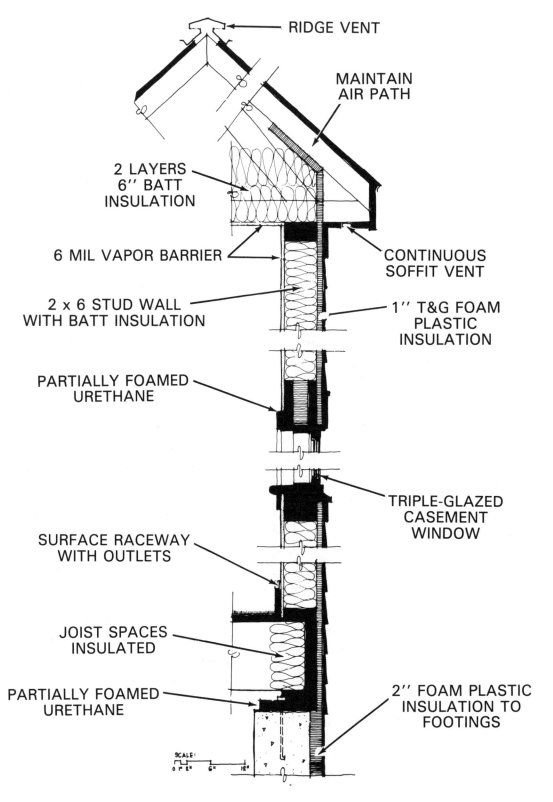

RIDGE VENT

MAINTAIN
AIR PATH

2 LAYERS
6'' BATT
INSULATION

6 MIL VAPOR BARRIER

CONTINUOUS
SOFFIT VENT

2 x 6 STUD WALL
WITH BATT INSULATION

1'' T&G FOAM
PLASTIC
INSULATION

PARTIALLY FOAMED
URETHANE

TRIPLE-GLAZED
CASEMENT
WINDOW

SURFACE RACEWAY
WITH OUTLETS

JOIST SPACES
INSULATED

PARTIALLY FOAMED
URETHANE

2'' FOAM PLASTIC
INSULATION TO
FOOTINGS

SCALE:

Unit 24 Quiz

Passive Solar Construction

Name _____ Score _____

True-False

Circle T if the answer is True or F if the answer is False.

T F 1. Convection only occurs in fluids or gases.

T F 2. Heat gained from appliances or occupants of a home is referred to as residential heat.

T F 3. Passive solar systems require pumps or blowers to carry heated air or fluid to the place where it will be used or stored.

T F 4. A major disadvantage of direct gain systems is the wide range of heat fluctuation.

T F 5. A Trombe wall is a nonbearing wall.

T F 6. In a direct gain system, the sun's rays enter directly through the glazing and heat up a thermal mass.

T F 7. Water is an effective material for the storage of solar heat.

Multiple Choice

Choose the answer that correctly completes the statement. Write the corresponding letter in the space provided.

_____ 8. Heat can travel by _____.

A. radiation
B. convection
C. conduction
D. All of the above.

_____ 9. The Trombe wall is an example of a(n) _____ solar system.

A. direct gain
B. indirect gain
C. isolated gain
D. None of the above.

_____ 10. Most of the windows in solar construction are placed in the _____ wall.

A. north
B. south
C. east
D. west

_____ 11. In the _____ gain system, the solar heat is collected and stored in an area away from the living or working space.

A. direct
B. indirect
C. isolated
D. None of the above.

_____ 12. A material commonly used for thermal storage wall construction is _____.
   A. water
   B. concrete
   C. concrete block
   D. All of the above.

330

25

Remodeling, Renovating, and Repairing

OBJECTIVES

Students will be able to:

- ❏ Distinguish between different types of residential construction and determine type by visual inspection.
- ❏ Set up a proper sequence of renovation and/or repair for the interior and exterior of a house.
- ❏ Repair and replace deteriorated components and systems.
- ❏ Remove parts of the structure without damaging the total structure.
- ❏ List steps for removal of wall sections prior to remodeling.
- ❏ Identify bearing walls.
- ❏ Determine loads and translate this information to correct header size for the span.
- ❏ Install and support headers, concealed headers, and saddle beams.
- ❏ Follow accepted methods in replacing all types of doors.
- ❏ Make repairs to wood and asphalt shingles.
- ❏ Make a solar retrofit on an older home.

INSTRUCTIONAL MATERIALS

Text: Pages 643-665
    Important Terms, page 664
    Test Your Knowledge, page 664
    Outside Assignments, page 664
Workbook: Pages 171-178
Instructor's Manual:
    Reproducible Master 25-1, *Remodeling Post-and-Beam Construction*
    Reproducible Master 25-2, *Replacing Rotted Sills*
    Unit 25 Quiz
    Section V Exam

TRADE-RELATED MATH

The method used to compute the load on a header is shown in Figure 25-24. Instruct students that the number of square feet must be determined first. The length of the span needed is then multiplied by half the distance between load-bearing walls. Any other loads that the header must support are added to find the total load.

INSTRUCTIONAL CONCEPTS AND STUDENT LEARNING EXPERIENCES

EXTERIOR RENOVATION

1. Instruct students about the importance of making structural repairs before attempting any other exterior renovation projects. Jacking of a sagging frame may affect fit of doors and windows; other repairs may be needed to ensure personal safety.
2. Review the sequence of tasks on page 644 for exterior renovation.
3. If you have "live" renovation work, have students examine the structure from the outside and prepare a list of the work that needs to be done. Have a class discussion in which students compare their lists and prepare a final work order for the renovation.
4. If "live" work is not available, secure photos of local structures and have students study them and list needed repairs and/or renovation tasks.

INTERIOR RENOVATION

1. Explain why interior renovations may coincide with exterior work unless the danger of leaks could ruin the interior work.
2. Discuss ways of doing interior renovation while the dwelling is still occupied. Review "zone by zone" and "room by room" renovation methods.
3. Review sequence for interior renovation.
4. If "live" work is available, have students prepare a sequence for the work to be done. Review in class and finalize the list before beginning the renovation.
5. If "live" work is not available, give students a

simulated work order for interior renovation and have them prepare a sequence of tasks.

REPLACING ROTTING SILLS

1. Review Figure 25-11 with students and discuss the method of supporting a building frame while a rotted member is replaced.
2. Explain "sistering" to support rotting vertical framing pieces.

DESIGN OF OLD STRUCTURES

1. Explain that many older structures used balloon framing or post-and-beam construction. Using Figure 25-9, review the characteristics of balloon framing. Include the "building height" studs, firestops, and let-in bracing.
2. Using Reproducible Master 25-1, *Remodeling Post-and-Beam Construction,* review the components of this type of structure. Emphasize to the students that they should be aware of the components and type of construction so they can anticipate possible problems.

HIDDEN STRUCTURE DETAILS

1. Discuss the type of inspection that should take place before any remodeling work begins. Have students identify the types of utilities they should be looking for, including gas or fuel lines, electrical circuits, and plumbing service. Stress that the utility companies should be notified if interruption to any of these services is anticipated.
2. Explain why an exterior inspection is necessary, and what they should be looking for when making an exterior inspection.

REMOVING OLD WALLS

1. Review the tools that are commonly used in remodeling work.
2. Discuss the preliminary steps in remodeling work, including the removal of trim and wall covering. Mention that, in many cases, trim members may need to be reused, so care should be taken in removing these items.
3. Stress the importance of safe work habits in remodeling work. Suggest that a large trash container be rented for major remodeling projects.

RECOGNIZING BEARING WALLS

1. Identify the characteristics of bearing walls. Stress that bearing walls will need temporary shoring while the old wall is being removed.

PROVIDING SHORING

1. Identify two types of temporary shoring that can be used for remodeling project. Emphasize that shoring should be placed on both sides of an interior bearing wall to prevent sagging.
2. Describe the construction of a short wall consisting of 2×4 plates and studs. Demonstrate how this short wall is constructed.
3. Discuss the use of adjustable steel posts for temporary shoring. Stress that the shoring must rest on adequate support, and that it should be positioned across several joists to distribute the load. Mention that if the shoring is running parallel to the floor joists that planking should be laid down to distribute the load.

FRAMING OPENINGS IN A BEARING WALL

1. Discuss the procedure used to frame openings in a bearing wall. Suggest possible methods of construction for the header.
2. Describe how a header is installed and supported.
3. Have students refer to the local building code to determine the size of header needed for a remodeling project.
4. Show the students how to compute the load on a header by using Figure 25-24. Emphasize that the load calculations should be checked (or specified) by an architect or structural engineer.
5. Discuss the use of concealed headers and saddle beams in remodeling. Describe the construction of each.

CONCEALED HEADERS AN SADDLE BEAMS

1. Refer to Unit 9 discussion of flush beams and strongbacks, as well as Figure 25-25, as you explain the construction of concealed headers and saddle beams.

REPLACING AN OUTSIDE DOOR

1. Refer to Figure 25-28 as you discuss replacement of a worn outside door with a steel unit.
2. Explain step-by-step how to install the door. Have students study Figure 25-29.
3. If "live" work is available, have students replace an outside door under your supervision.

REPLACING OR REPAIRING AN INTERIOR DOOR

1. Have students review Unit 9 on framing a door opening and Unit 18 for hanging an interior door. Refer them to Figures 25-30 through 25-34.

2. Review steps for correcting problems such as binding or sagging doors.
3. Demonstrate method for trimming the bottom of an existing door to accommodate new flooring materials.

INSTALLING NEW WINDOWS

1. Demonstrate steps for removing a window that is to be replaced. Explain that removal of trim must be done carefully since it may be necessary to use it.
2. Demonstrate installation and fastening of a replacement window. Explain how to reframe the rough opening if the new window is not the same size.

REPAIRING WOOD AND ASPHALT SHINGLES

1. Explain and demonstrate the method of removing and replacing damaged wood shingles.
2. Explain and demonstrate methods for replacing damaged asphalt shingles.

BUILDING ADDITIONS ONTO OLDER HOMES

1. Discuss the items that should be checked when building additions onto older homes. Stress that dimensions for framing members, ceiling heights, and foundations may have been different when the structure was originally built. Explain what should be done if the dimensions vary a great deal.

SOLAR RETROFIT

1. Review the three passive solar designs—direct gain, indirect gain, and isolated space.
2. Describe the work involved in installing a direct gain system in an existing structure.
3. Discuss two methods of installing a thermosiphon (solar furnace).

RESPONSIBLE RENOVATION

1. Explain the concept of recycling of old building materials where possible.
2. Have students study the interior and exterior of a remodeling project to see what might be recyclable.
3. Visit a recycling center to see what might be done with recycled building materials.

FALL PROTECTION

1. Procure a copy of OSHA guidelines on fall protection from a local builder and discuss with your students.

2. Review photos of scaffolding or study on-site scaffolding to see what measures are taken to prevent falls.

UNIT REVIEW

1. Review the unit objectives. Be sure that the students fully understand each objective.
2. Assign *Important Terms, Test Your Knowledge* questions, and *Outside Assignments* on page 684 of the text. Review the answers in class.
3. Assign pages 171-178 of the workbook. Review the answers in class.

EVALUATION

1. Use Unit 25 Quiz for in-class evaluation. Correct the quizzes and return them to the students for review.
2. Use the Section 5 Exam to evaluate the student's knowledge of information found in Units 21-25 of the text.

ANSWERS TO TEST YOUR KNOWLEDGE TEXT PAGE 664

1. D. Repairing any leaks found.
2. List what needs to be done.
3. paint
4. Other activities might damage the finish on the floor.
5. B. Building must be supported some way to remove load from sills.
6. Placing another framing member alongside a damaged one to strengthen it.
7. It will help you in dismantling sections of the house.
8. pipes or cables, buried
9. A. The wall runs down the middle of the length of the house.
   B. Overhead joists are spliced over the wall indicating they depend on the wall for support.
   C. The wall runs at right angles to overhead joists and breaks up a long span.
   D. Wall is directly below a parallel wall on the upper level.
10. Shoring
11. 8″
12. Saddle beam or concealed header.
13. True.
14. isolated
15. Fitting an existing building with structures that make use of solar energy for conditioning the building interior space.

16. It is friendlier to the environment and saves natural resources.

ANSWERS TO WORKBOOK QUESTIONS PAGES 171-178

1. D. Perform all structural work, proceeding from the ground up.
   H. Regrade site and provide drainage away from house.
   C. Repair or replace the roof.
   A. Repair or replace windows.
   B. Repair or replace siding.
   E. Stain or prime wood siding.
   G. Caulk, glaze, and putty.
   F. Paint.
2. Jacks have been placed to raise and support house frame in preparation for replacing a rotten sill.
3. visualize
4. A. Ribbon/ledger.
   B. Joist.
   C. Firestopping.
   D. Sill.
   E. Firestopping
   F. Stud.
5. underground
6. B. garden spade
7. B. The wall runs parallel to an outside wall
8. (See Figure 25-17 in the text.)
9. A. Rafter.
   B. Joist.
   C. Firestopping.
   D. Subflooring.
   E. Sill
   F. (Let in) brace.
   G. Stud.
10. trimmer studs
11. A. 20 lb.
    B. 30 lb.
12. joist hangers
13. larger
14. C. 8d box
15. A. Remove old door hinges, strike plate, threshold, and frame.
16. C. Use magnetic type on lock side and top.
17. B. Fasten clips to door jamb and then install the assembly in the rough opening.
18. C. 150 lb.
19. thermosiphon
20. True.

ANSWERS TO UNIT 25 QUIZ

1. B. doing temporary repairs to stop further deterioration
2. D. tear out all that is being eliminated and remove debris
3. A. remove all trim
4. C. trimmers
5. B. solar furnace
6. G
7. D
8. H
9. F
10. E
11. A
12. J
13. I
14. B
15. C

ANSWERS TO SECTION 5 EXAM

1. False.
2. True.
3. False.
4. True.
5. True.
6. False.
7. True.
8. False.
9. Longitudinal
10. B. Stressed skin panels
11. D. All of the above.
12. A. internal heat
13. D. All of the above.
14. D. All of the above.
15. B
16. E
17. D
18. A
19. C
20. thermosiphon
21. isolated
22. trimmer
23. Jack posts
24. Trombe wall
25. south
26. Convection
27. 2, 4
28. chase
29. Underwriters' Laboratories

30. G
31. A
32. C
33. E
34. F
35. B
36. D
37. E
38. A
39. D
40. C
41. F
42. B

Reproducible Master 25-1
Remodeling Post-and-Beam Construction

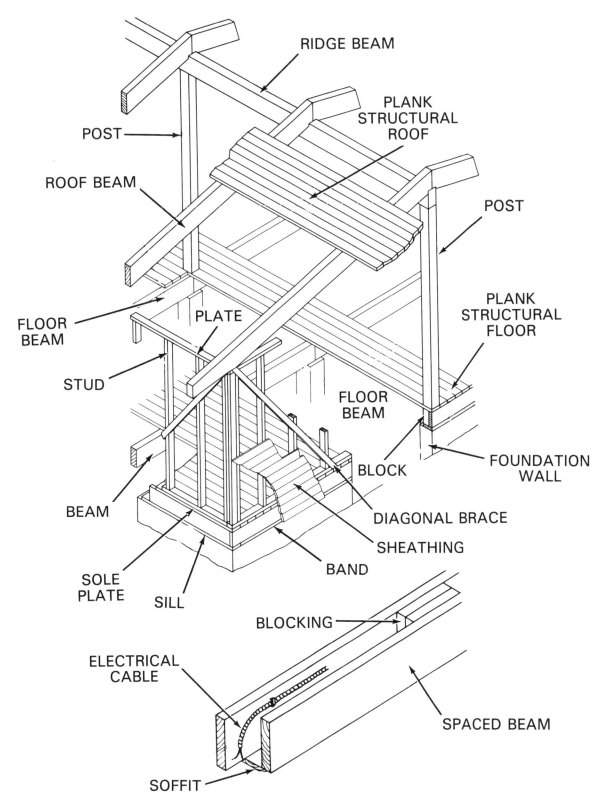

RIDGE BEAM

PLANK STRUCTURAL ROOF

POST

ROOF BEAM

POST

PLANK STRUCTURAL FLOOR

FLOOR BEAM

PLATE

STUD

FLOOR BEAM

FOUNDATION WALL

BEAM

BLOCK

DIAGONAL BRACE

SOLE PLATE

SILL

SHEATHING

BAND

BLOCKING

ELECTRICAL CABLE

SPACED BEAM

SOFFIT

Reproducible Master 25-2
Replacing Rotted Sills

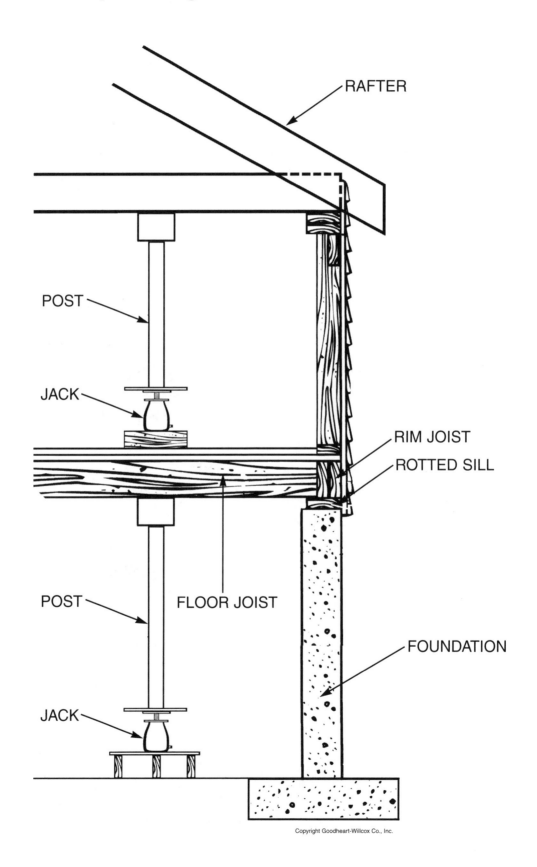

RAFTER

POST

JACK

RIM JOIST

ROTTED SILL

POST

FLOOR JOIST

FOUNDATION

JACK

Unit 25 Quiz

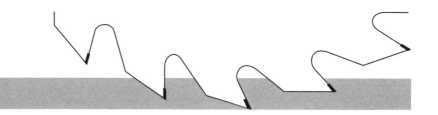

Remodeling, Renovating, and Repairing

Name _____ Score _____

Multiple Choice

Choose the answer that correctly completes the statement. Write the corresponding letter in the space provided.

_____ 1. A first step in repair, remodeling, or renovation of a long-neglected house should be _____ .

      A. painting the exterior
      B. doing temporary repairs to stop further deterioration
      C. demolishing "add-ons" that are to be removed
      D. making needed foundation repairs

_____ 2. When making interior renovations, first _____.

      A. make alterations affecting walls–especially bearing walls
      B. rough in plumbing and electrical changes
      C. insulate exterior walls
      D. tear out all that is being eliminated and remove the debris

_____ 3. The first step in opening up a wall is to _____.

      A. remove all trim
      B. remove headers
      C. cut studs using a reciprocating saw
      D. remove the wall covering

_____ 4. Headers are supported by resting them on top of short studs called _____.

      A. joists
      B. girders
      C. trimmers
      D. None of the above.

_____ 5. A thermosiphon is also referred to as a _____.

      A. direct gain system
      B. solar furnace
      C. thermostat
      D. None of the above.

Identification

Identify the parts of the balloon framed structure.

_____ 6. Sill.

_____ 7. Girder.

_____ 8. Sheathing.

_____ 9. Subflooring.

_____ 10. Ledger.

_____ 11. Firestopping.

_____ 12. Let-in brace.

_____ 13. Ribbon.

_____ 14. Stud.

_____ 15. Joist.

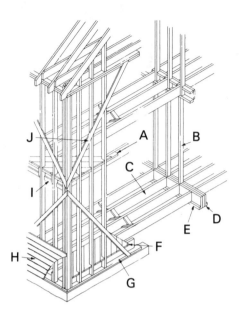

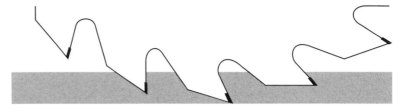

Section 5 Exam
Special Construction Techniques

Name _____ Score _____

True-False

Circle T if the answer is True or F if the answer is False.

T F 1. Footings that are 2″ thick are sufficient for masonry chimneys.

T F 2. Longitudinal beams are also called purlin beams.

T F 3. In panelized prefabrication, entire sections of the structure are built and finished in manufacturing plants.

T F 4. Heat always travels from a higher temperature to a lower temperature.

T F 5. One major disadvantage of direct gain solar systems is the wide range of heat fluctuations.

T F 6. The amount of overhang used for solar structures in the southern United States should be greater than the amount used in northern parts of the country.

T F 7. Thicker storage walls in a passive solar structure will store more heat than thinner walls.

T F 8. When remodeling a structure, shoring need only be placed on one side of a bearing wall.

Multiple Choice

Choose the answer that correctly completes the statement. Write the corresponding letter in the space provided.

_____ 9. _____ beams run parallel to the supporting side walls and ridge beam in post-and-beam construction.

    A. Transverse
    B. Steel
    C. Longitudinal
    D. All of the above.

_____ 10. _____ are made by gluing sheets of plywood to longitudinal framing members or other core materials.

    A. Longitudinal beams
    B. Stressed skin panels
    C. Transverse beams
    D. Box beams

_____ 11. The basic type(s) of passive solar energy is/are _____.

    A. indirect gain
    B. direct gain
    C. isolated gain
    D. All of the above.

341

_____ 12. Heat that is gained from household appliances and occupants is called _____.

    A. internal heat
    B. radiation
    C. solar energy
    D. None of the above.

_____ 13. Which of the following usually indicates a bearing wall?

    A. The wall runs down the middle of the length of the house.
    B. Overhead joists are spliced over the wall.
    C. The wall runs at right angles to overhead joists and breaks up a long span.
    D. All of the above.

_____ 14. Beam sizes for post-and-beam construction are based on the _____.

    A. span
    B. deflection permitted
    C. load they must carry
    D. All of the above.

Matching

Select the correct answer from the list on the right and place the corresponding letter in the blank on the left.

_____ 15. Hearth.

_____ 16. Damper.

_____ 17. Smoke shelf.

_____ 18. Smoke chamber.

_____ 19. Flue.

    A. Space extending from the top of the throat to the bottom of the flue.
    B. Consists of two parts: one located in the front and the other under the fire area.
    C. Conveys the smoke out of the fireplace.
    D. Helps to change the flow direction of the downdraft.
    E. Used to control the fire intensity.

Completion

Write the answer that correctly completes the statement in the space provided.

_____ 20. A(n) _____ is also referred to as a solar furnace.

_____ 21. In a(n) _____ solar system, there are separate spaces that have solar storage systems for collecting and distributing heat to other parts of the structure.

_____ 22. Headers are supported on top of short studs called _____ studs.

_____ 23. _____ _____ are adjustable steel pillars used to support ceiling structures when remodeling.

_____ 24. A(n) _____ _____ is a thick masonry wall placed next to the exterior glazing to store solar energy in passive solar construction.

_____ 25. Most of the windows in a solar structure should be placed in the _____ (direction) wall so that solar radiation can be collected.

_____ 26. _____ only occurs in fluids and gases.

_____ 27. Planks for floor and roof decking in post-and-beam construction may range in size from _____ to _____ inches thick, depending on the span.

Name _____

_____ 28. A(n) _____ is a wood frame extending from an outside wall which supports a prefabricated chimney.

_____ 29. A prefabricated chimney should have a label that indicates it has been tested by the _____ or another recognized testing organization.

Identification

Identify the parts of the sill construction for post-and-beam framing.

_____ 30. Sole plate.

_____ 31. Post.

_____ 32. Beam.

_____ 33. Foundation.

_____ 34. Header.

_____ 35. Plank.

_____ 36. Sill.

Identification

Identify the indicated parts of the fireplace cross section.

_____ 37. Throat.

_____ 38. Chimney flue.

_____ 39. Smoke shelf.

_____ 40. Smoke chamber.

_____ 41. Fireplace.

_____ 42. Flue lining.

26

Electrical Systems

OBJECTIVES

Students will be able to:

❏ Define basic electrical terms.
❏ Explain what is included in an electrical wiring system.
❏ List the tools, devices, and materials required to do electrical wiring in a residential building.
❏ Demonstrate the proper use of tools and handling of materials.
❏ Demonstrate an understanding of basic circuit theory.
❏ Use approved methods for simple wiring installation tasks.
❏ With minimal supervision, perform simple electrical troubleshooting.

INSTRUCTIONAL MATERIALS

Text: Pages 669-682
    Important Terms, page 681
    Test Your Knowledge, page 681
    Outside Assignments, page 682
Workbook: Pages 179-182
Instructor's Manual:
    Reproducible Master 26-1, *Electrical Symbols*
    Unit 26 Quiz

INSTRUCTIONAL CONCEPTS AND STUDENT LEARNING EXPERIENCES

INTRODUCTION

1. Define electrical wiring and explain what is included.
2. Explain current and its laws.
3. Introduce the National Electrical Code and explain its importance to the electrical trades.

TOOLS, EQUIPMENT, AND MATERIALS

1. Discuss the tools used in the electrical trade and explain their purposes.

2. Demonstrate, if possible, the use of each tool.
3. If they are available, show conductors, boxes, fuses, circuit breakers, switches, receptacles, conduit, and connectors. Explain the purpose of each.

BASIC ELECTRICAL WIRING THEORY

1. Explain the theory of electron flow through conducting material.
2. Note the difference between alternating current and direct current.
3. Discuss transformers, explaining their purpose and construction.
4. Note the difference between a step-up transformer and a step-down transformer.
5. Explain the two levels of voltage used in residential wiring.

INSTALLING THE SERVICE

1. Define the terms, "service" and "distribution panel."
2. Explain where the service is found, where it begins, and where it ends.
3. List all the devices found inside a distribution panel.

READING BLUEPRINTS

1. Using Reproducible Master 26-1, *Electrical Symbols*, go over the symbols commonly found in a home's blueprint. Explain that these symbols are placed on house plans to show where each device is to be located.
2. Wherever possible, show each device along with its symbol.
3. Obtain a blueprint showing a wiring plan and pass it among the students for study.
4. Discuss the electrician's use of the blueprint in installing an electrical system in a residence.

RUNNING BRANCH CIRCUITS

1. Define branch circuitry.

2. Refer again to the definition of a circuit as a path along which electric current may travel.

3. Discuss the types of control devices and current-using devices that are part of a branch circuit.

4. Review the types of devices used to protect electrical conductors from damage.

DEVICE WIRING

1. Demonstrate the proper technique for making electrical connections of two conductors and conductors and devices.

2. Demonstrate stripping of insulation from the ends of conductors.

3. Wire up a simple circuit and demonstrate that it works when plugged into a power source.

4. Referring to Figure 26-29, explain the operation of a three-way switch.

ELECTRICAL TROUBLESHOOTING

1. Refer to Figures 26-30 through 26-34 as you demonstrate test procedures for testing receptacles, switches, and fixtures.

UNIT REVIEW

1. Review the Unit Objectives. Be sure that the students fully understand each objective.

2. *Assign Important Terms, Test Your Knowledge* questions, and *Outside Assignments* on pages 681 and 682 of the text. Review the answers in class.

3. Assign pages 179-182 of the workbook. Review the answers in class.

EVALUATION

1. Use Unit 26 Quiz for in-class evaluation. Correct the quizzes and return them to the students for review.

ANSWERS TO TEST YOUR KNOWLEDGE
TEXT PAGE 681

1. Wires, boxes, and a number of devices that control the distribution and use of electrical current.

2. The flow of electrons through a conducting material.

3. National Electrical Code

4. For boring holes in studs and joists where electrical wiring is to run.

5. No. It would be a hazard since it conducts electricity.

6. Switches
7. alternating
8. True.
9. False.
10. A complete path along which electric current can travel.
11. In the circuit conductors.
12. See Figure 26-34.

ANSWERS TO WORKBOOK QUESTIONS
PAGES 179-182

1. before
2. False.
3. fish tape
4. A. wires
   B. current
5. A. Circuit breaker, single pole.
   B. Load center, distribution panel, or entrance panel (any of the three)
   C. Meter box.
   D. Single pole switch.
   E. Receptacle or convenience outlet (either one).
6. A. Fuses
   B. circuit breakers
7. False.
8. service
9. A. Ceiling outlets for fixtures.
   B. Wall fixture outlet.
   C. Outside telephone connection.
10. A. hot or black wire
11. It is a test to see if the circuit is live.
12. False. (It could also mean that the circuit breaker is tripped or that one of the conductors is at fault.)

ANSWERS TO UNIT 26 QUIZ

1. False.
2. False.
3. True.
4. False.
5. D
6. B
7. E
8. A
9. C
10. D. switch
11. A. voltage
12. C. distribution panel

Electrical Symbols

ELECTRICAL SYMBOLS	
CEILING OUTLETS FOR FIXTURES	▬ LIGHTING PANEL
	▨ POWER PANEL
WALL FIXTURE OUTLET	S SINGLE-POLE SWITCH
CEILING OUTLET WITH PULL SWITCH	S_2 DOUBLE-POLE SWITCH
WALL OUTLET WITH PULL SWITCH	S_3 THREE-WAY SWITCH
DUPLEX CONVENIENCE OUTLET	S_4 FOUR-WAY SWITCH
WATERPROOF CONVENIENCE OUTLET	S_P SWITCH WITH PILOT LIGHT
CONVENIENCE OUTLET 1, 3, 1 = SINGLE 3 = TRIPLE	⊡ PUSH BUTTON
RANGE OUTLET	⬭) BELL
CONVENIENCE OUTLET WITH SWITCH	◀ OUTSIDE TELEPHONE CONNECTION
SPECIAL PURPOSE (SEE SPECS.)	TV TELEVISION CONNECTION
FLOOR OUTLET	S SWITCH WIRING
CEILING LIGHT FIXTURE	▭ EXTERIOR CEILING FIXTURE
PULL CHAIN LIGHT FIXTURE	FLUORESCENT CEILING FIXTURE
EXTERIOR LIGHT FIXTURE	FLUORESCENT WALL FIXTURE

Unit 26 Quiz

Electrical Systems

Name _____ Score _____

True-False

Circle T if the answer is True or F if the answer is False.

T   F   1. The National Electrical Code is a collection of laws on how electrical systems must be installed.

T   F   2. Wire connectors are slowly being replaced by soldered electrical connections.

T   F   3. Fiberglass ladders are recommended for electrical work because they are not conductors of electricity.

T   F   4. Most homes use direct current instead of alternating current.

Identification

Identify the electrical symbols.

_____   5. Single pole switch.

_____   6. Duplex convenience outlet.

_____   7. Double pole switch.

_____   8. Ceiling outlets for fixtures.

_____   9. Floor outlet.

Multiple Choice

Choose the answer that correctly completes the statement. Write the corresponding letter in the space provided

_____   10. Manual control of circuits is possible through installation of a(n) _____ in the "hot" conductor.
     A. fuse
     B. circuit breaker
     C. receptacle
     D. switch

_____   11. A transformer changes _____ in an electrical system, utilizing principles of induced magnetism.
     A. voltage
     B. current
     C. resistance
     D. power

_____   12. A box that houses a main breaker and circuit breakers or fuses for all house circuits is called a(n) _____.
     A. fuse box.
     B. breaker box.
     C. distribution panel
     D. service

27

Plumbing Systems

OBJECTIVES

Students will be able to:

- ❏ Cite codes that govern the installation of plumbing systems.
- ❏ List necessary plumbing tools and explain how to use them.
- ❏ Describe the different types of materials used in plumbing systems.
- ❏ Name and recognize devices and fixtures that are part of the plumbing system.
- ❏ Explain the proper design and installation of plumbing systems.
- ❏ Understand plumbing prints.
- ❏ Demonstrate a basic understanding of well and pump systems.
- ❏ Unplug drains using chemicals and the special tools designed to clear out clogs.
- ❏ Cite safety measures that plumbers need to observe.

INSTRUCTIONAL MATERIALS

Text: Pages 683-700
  Important Terms, 699
  Test Your Knowledge, page 700
  Outside Assignments, page 700
Workbook: Pages 183-184
Instructor's Manual:
  Reproducible Master 27-1, *Typical Plumbing System*
  Reproducible Master 27-2, *Parts of a Globe Valve*
  Unit 27 Quiz

TRADE-RELATED MATH

Often, plumbers must compute the lengths of pipe offsets. One method is to use a formula designed to find one side of a right-angle triangle. The theory behind the formula is that the square of the hypotenuse (side opposite a 90° angle) of a right-angle triangle is equal to the sum of the squares of the other two sides. For example, suppose that the two pipes to be joined are 10″ apart and one is 12″ above the other:

$$10^2 + 12^2 = H$$
$$100 + 144 = 244$$
$$\sqrt{244} = 15.62″$$

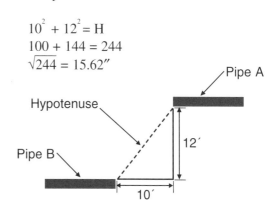

INSTRUCTIONAL CONCEPTS AND STUDENT LEARNING EXPERIENCES

PLUMBING CODES

1. Discuss the necessity of maintaining standards and the possible repercussions if these standards are not followed.
2. Distribute copies of your community's plumbing code to your students. Discuss and compare the code with that of another community.

TWO SEPARATE SYSTEMS

1. Using Reproducible Master, 27-1, *Typical Plumbing System*, compare the design and functions of the two separate parts of a plumbing system.
2. Explain why one is pressurized and the other works by gravity flow.
3. Stress the importance of making both systems watertight.
4. Discuss the purpose of venting and explain how it is designed.

TOOLS

1. Display a full kit of plumbing tools, if available, and explain how to use them.

MATERIALS

1. List the various types of materials that make up the plumbing system. Explain where they might be used.
2. Demonstrate how to make connections with each type of material.
3. Have students practice making connections.

VALVES

1. Discuss the function of a valve in a pressurized supply system.
2. Using Reproducible Master 27-2, *Parts of a Globe Valve*, review the construction and parts of a globe valve. Remind students that there are other designs for valves that are in common use.

FAUCETS

1. Compare the function and construction of faucets with valves.

FIXTURES

1. Define the term *fixture*.
2. Have students list the different types of fixtures found in a modern residence.

PRINTREADING

1. Have students review Unit 6.
2. Discuss the use of symbols to represent plumbing devices and fixtures. Refer to Figures 27-15 and 27-16.
3. Have students draw a plumbing system for a small house. Discuss the designs in class.

INSTALLING PLUMBING

1. Point out that plumbing is usually done by specialists, but that a skilled homeowner can do simple installations or remodeling.
2. Demonstrate the cutting of pipes and tubing with both saws and cutters.
3. Have students practice cutting and reaming operations.

SWEAT SOLDERING

1. Demonstrate the proper procedure for sweat soldering. Impress upon your students the necessity for starting with a clean surface that has been brightened with steel wool.

2. Explain the role of flux in producing a leak-free joint.
3. If time permits, allow students to practice sweat soldering a joint.

MAKING COMPRESSION JOINTS

1. Explain the principle of compression joints.
2. Demonstrate the procedure for completing a compression joint.

BENDING AND UNROLLING COPPER TUBING

1. Explain method of bending and unrolling copper tubing.

MAKING PLASTIC PIPE CONNECTIONS

1. Demonstrate cleaning, applying adhesive and making the fitting connection.
2. Caution students that speed and accuracy is extremely important as the adhesive will set up in seconds, making any adjustment impossible.

MAKING GALVANIZED PIPE CONNECTIONS

1. Mention that this is one of the easiest connections to make and change.
2. Explain that these pipe connections should always be made using either pipe compound or teflon tape.
3. Demonstrate procedure and then have students make connections.

WELLS AND PUMPS

1. Discuss the types of wells. Mention that most wells today are drilled or bored.
2. List the various essential parts of the system and explain the role of each.

UNPLUGGING DRAINS

1. Demonstrate the use of various tools commonly used to unplug drains.

SAFETY

1. Discuss safety rules and proper attire.

UNIT REVIEW

1. Review the unit objectives. Be sure that students fully understand each objective.
2. Assign *Important Terms, Test Your Knowledge* questions, and *Outside Assignments* on pages 699 and 700. Review the answers in class.
3. Assign pages 183-184 of the workbook. Review the answers in class.

EVALUATION

1. Use Unit 27 Quiz for in-class evaluation. Correct the quizzes and return them to the students for review.

ANSWERS TO TEST YOUR KNOWLEDGE TEXT PAGE 700

1. Indiscriminate notching and other alterations to a building's frame could weaken the structure.
2. An improperly designed system will not pass inspection and would have to be redone.
   A poorly designed system will not perform well.
   A poorly designed system may cause health problems through contaminants entering the supply water.
3. water supply system
4. Stands for "drainage, waste, and venting." It carries away waste water.
5. level, plumb
6. Supply pipe: copper, galvanized steel, and plastic. DWV: malleable iron, copper, galvanized steel, and plastic.
7. Stands for Chlorinated Polyvinyl Chloride and can be used for piping hot water lines.
8. A fixture is a water-consuming device in a plumbing system. Examples include laundry tubs, lavatories, sinks, showers, and bath tubs.
9. On copper pipes to make leakproof connections.
10. With a special adhesive.
11. rough-in
12. Well and well casing, pump, and supply tank.
13. Chemical cleaners, plungers, snakes, and closet augers.
14. See Figure 27-27 and related text.

ANSWERS TO WORKBOOK QUESTIONS PAGES 183-184

1. Uniform Plumbing Code.

BOCA Basic Building Code.
ICBO Plumbing Code.
National Plumbing Code.
Standard Plumbing Code.
2. Venting
3. C. Malleable iron.
4. CVPC, PB
5. A. Hub and spigot.
   B. No-hub.
6. A. Stem.
   B. Packing nut.
   C. Valve seat.
   D. Bonnet.
7. fixtures
8. False.
9. True.
10. tubing spring
11. Drilled
   Bored
12. Seals harden and leak.
   Check valves wear out.
   Pumps wear and lose efficiency or stop working.
   Galvanized pipes rust and leak.
13. Molten lead is poured into the joint.
   A neoprene gasket is installed on the joint.

ANSWERS TO UNIT 27 QUIZ

1. False.
2. False.
3. True.
4. False.
5. A
6. C
7. B
8. C. sweat soldering (compression joints)
9. E. All of the above.
10. B. supply tank

Typical Plumbing System

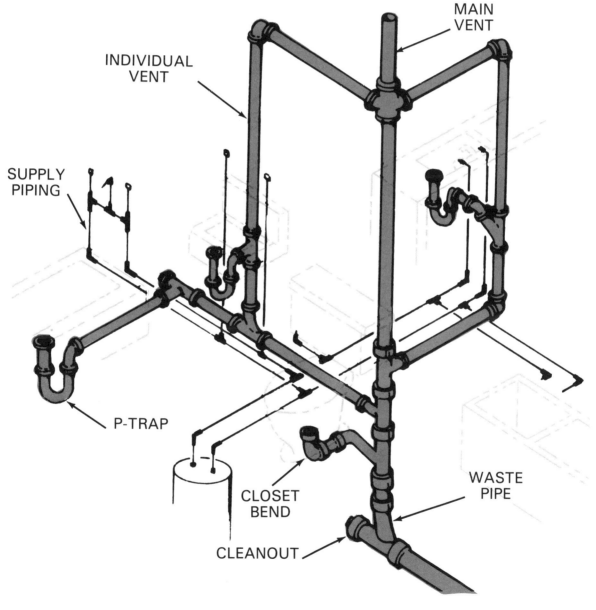

MAIN VENT

INDIVIDUAL VENT

SUPPLY PIPING

P-TRAP

CLOSET BEND

CLEANOUT

WASTE PIPE

Reproducible Master 27-2
Parts of a Globe Valve

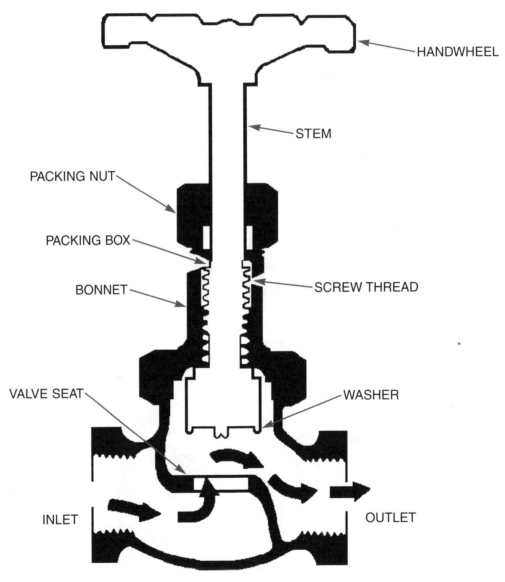

HANDWHEEL

STEM

PACKING NUT

PACKING BOX

SCREW THREAD

BONNET

VALVE SEAT

WASHER

INLET

OUTLET

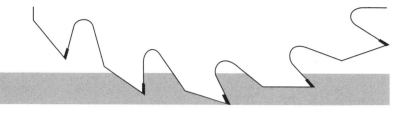

Unit 27 Quiz

Plumbing Systems

Name _____ Score _____

True-False

Circle T if the answer is True or F if the answer is False.

T   F   1. Plumbers are not allowed to cut or notch framing members; they must have carpenters perform these tasks.

T   F   2. Since leveling of plumbing is not necessary, a level would not be an important plumbing tool.

T   F   3. CVCP connections can be threaded or solvent welded.

T   F   4. A tubing spring is permanently installed at bends in copper tubing

Identification

Identify the types of plumbing sketches in the illustrations.

_____   5. Plan view.

_____   6. Riser diagram.

_____   7. Isometric sketch.

A
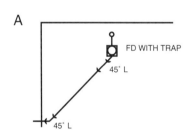

B
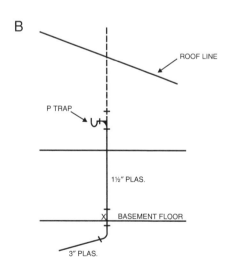

C
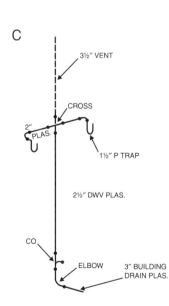

Multiple choice

Choose the answer that correctly completes the statement. Write the corresponding letter(s) in the space provided.

_____ 8. Copper plumbing connections are made by _____.

    A. solvent bonding
    B. welding
    C. sweat soldering
    D. None of the above.

_____ 9. A _____ _____ is used for unclogging a drain.

    A. drain cleaner
    B. plunger
    C. snake
    D. closet auger
    E. All of the above.

_____ 10. A well system uses a _____ _____ to store water under pressure so it will available upon demand.

    A. well casing
    B. supply tank
    C. submersible pump
    D. check valve

28

Heating, Ventilation, and Air Conditioning

OBJECTIVES

Students will be able to:

- ❏ Suggest ways of conserving energy in housing.
- ❏ Define EER ratings and list the appliances to which they are applied.
- ❏ List and give the characteristics of different central air conditioning systems.
- ❏ Name and give the function of system components.
- ❏ Detail the design and operation of these systems.
- ❏ Discuss automatic controls for heating and cooling systems.

INSTRUCTIONAL MATERIALS

Text: Pages 701-715

    Important Terms, page 712

    Test Your Knowledge, page 713

    Outside Assignments, page 714

Workbook: Pages 185-188

Instructor's Manual:

    Reproducible Master 28-1, *Fundamentals of a Gas Furnace*

    Unit 28 Quiz

    Section 6 Exam

TRADE-RELATED MATH

Basic math skills are essential to the person installing heating and cooling appliances. These skills are needed for making accurate measurements, properly replacing parts and completing work orders and time sheets.

Example: A part-time hourly worker worked 6 hours on Monday, 8 hours on Tuesday. At a pay rate of $7.50 per hour, what amount will he or she be paid for this work?

    6 hours + 8 hours = 14 hours

    14 hours × $7.50/hr. = $105

INSTRUCTIONAL CONCEPTS AND STUDENT LEARNING EXPERIENCES

CONSERVATION MEASURES

1. Review the three methods by which heat is transported from one point to another.
2. Relate these physical principles to measures that will conserve heat energy in a building.
3. Point out to students that cost effectiveness of any conservation measure is directly related to the costs of conservation measures, and the cost of energy used to heat or cool a building. As fuel and energy costs rise, conservation becomes more cost effective.
4. Discuss the meaning of an EER and its significance in choosing between one brand of appliance over another.

CENTRAL HEATING SYSTEMS

1. Explain to students that heating systems make use of the physical principles governing the movement of heat from one location to another.
2. List and describe the four types of central heating systems.
3. Discuss the method of heat transfer used by each system.
4. Describe the role of the components of each system in controlling and moving heat.
5. Use Reproducible Master 28-1, *Fundamentals of a Gas Furnace*, to explain the sequence of events in a gas furnace. Begin with the thermostat calling for heat to the signal from the thermostat to stop sending heat.

AIR COOLING SYSTEMS

1. Discuss the type of transfer medium used by air conditioning systems.
2. Remind students that the principles of heat transfer also apply to appliances that cool air.

3. Compare the air conditioner to a refrigeration unit.
4. Explain the roles of the cooling coil and the condenser.

CONTROLS

1. Explain that controls react to sensing devices or electric signals that cause them to control operation of furnaces and air conditioners.
2. Explain the terms "cut-in point," "cut-out point," and "differential."
3. Explain the difference between series connection and parallel connection of the thermostat in the control circuit.

AIR EXCHANGERS

1. Explain the purpose of an air exchanger.
2. Discuss its operation.
3. Explain that with modern airtight homes, an air exchanger may be necessary to maintain sufficient fresh air in the building.

UNIT REVIEW

1. Review the unit objectives. Be sure that the students fully understand each objective.
2. Assign *Important Terms, Test Your Knowledge* questions, and *Outside Assignments* on pages 712-714 of the text. Review the answers in class.
3. Assign pages 185-188 of the workbook. Review the answers in class.

EVALUATION

1. Use Unit 28 Quiz for in-class evaluation. Correct the quizzes and return them to the students for review.
2. Use the Section 6 Exam to evaluate the students' knowledge of the information found in Units 26-28 of the text.

ANSWERS TO TEST YOUR KNOWLEDGE
TEXT PAGE 713

1. Sealing of cracks at joints with caulk and housewrap.
   Increasing the amount of insulation in walls, ceilings, and floors.
   Using an air exchanger to wring heat out of air being exhausted.
   Replacing heating and cooling appliances with newer ones having higher efficiency ratings.
2. False.
3. A heat exchanger is part of a heating unit that passes air heated by hot combustion gases to the plenum; an air exchanger wrings heat out of exhaust air and passes it along to fresh incoming air.
4. An EER is an "energy efficiency ratio"; it indicates how well an appliance uses energy. This rating covers furnaces, central and room air conditioners, as well as other major appliances.
5. Forced air perimeter heating system.
   Hydronic perimeter heating.
   Hydronic radiant heating.
   Electric resistance radiant heating in ceilings or walls.
6. Inspect panels and replace when caked with minerals from the water supply.
   Check for proper flow of water.
7. warm-air perimeter
8. Air is the medium in one while water is the medium in the other.
9. A separate pump for each zone delivers heated water when the zone thermostat calls for more heat.
10. single-pipe
11. It acts as a reservoir when heated water expands.
12. perimeter
13. False.
14. thermostat
15. heat pump
16. True.

ANSWERS TO WORKBOOK QUESTIONS
PAGES 185-188

1. A. it stops infiltration or exfiltration of inside conditioned air by way of convection currents
2. Evaluate individually. (Basically, chimneys account for a great deal of heat loss for two reasons: To draw well, chimneys first have to be heated to about 300°F; secondly, heat naturally rises and readily escapes via the chimney.
3. A. Heat exchanger.
   B. Burners.
   C. Blower or fan.
   D. Control system.
   E. Gas valve.
   F. Vent pipe.
4. A. Boiler.
5. air return
6. Any two of the following: sheet metal, vitrified tile, concrete pipe, plastic pipe.
7. D. the building has zone heating
8. A. Baseboard units.
   B. Return pipe.

C. Pump.

D. Supply pipe.

9. A. cooling coil

B. condenser

10. Window air conditioner.

11. C. every two to three hours

ANSWERS TO UNIT 28 QUIZ

1. True.
2. False.
3. True.
4. False.
5. Ducts
6. heating season
7. E
8. C
9. A
10. B
11. D

ANSWERS TO SECTION 6 EXAM

1. False.
2. True.
3. True.
4. False.
5. False.
6. False.
7. D. All of the above.
8. D. All of the above.
9. C. hydronic radiant

10. Conductors
11. perimeter
12. Oakum
13. kinks
14. chimneys
15. Registers
16. B
17. E
18. D
19. C
20. A
21. B
22. D
23. A
24. C
25. B
26. C
27. A
28. D
29. C
30. A
31. B
32. H
33. G
34. A
35. B
36. E
37. C
38. D
39. F

Fundamentals of a Gas Furnace

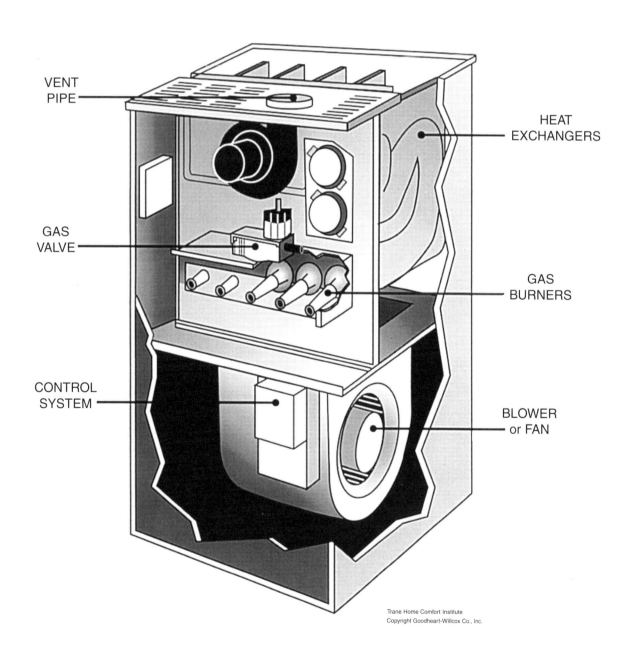

VENT
PIPE

HEAT
EXCHANGERS

GAS
VALVE

GAS
BURNERS

CONTROL
SYSTEM

BLOWER
or FAN

Trane Home Comfort Institute
Copyright Goodheart-Willcox Co., Inc.

Unit 28 Quiz

Heating, Ventilation, and Air Conditioning

Name _____ Score _____

True-False

Circle T if the answer is True or F if the answer is False.

T F   1. Four factors are involved in energy efficient homes: sealing cracks, insulation levels, wringing heat out of polluted warm air before exhausting it, and efficient appliances.

T F   2. Modern furnaces can achieve efficiencies up to 80%.

T F   3. Among other items, an EER label compares an appliance's efficiency with that of comparable products on the market.

T F   4. A heat exchanger wrings heat from exhausting warmed air.

Sentence completion

Complete each of the sentences with the correct word or phrase.

_____   5. _____ are passage ways that carry conditioned air from a heating or cooling appliance and distributes it to various rooms of a building.

_____   6. Blower belts on a forced air system should be inspected before every _____ _____.

Identification

Identify the indicated parts.

_____   7. Supply pipe.

_____   8. Pump.

_____   9. Baseboard units.

_____ 10. Return pipe.

_____ 11. Boiler

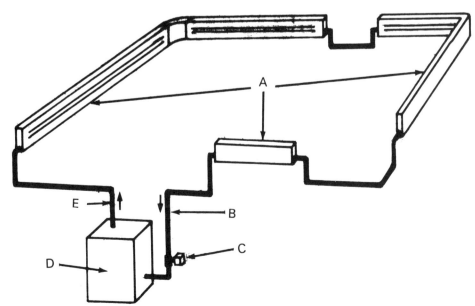

Section 6 Exam
Mechanical Systems

Name _____ Score _____

True-False

Circle T if the answer is True or F if the answer is False.

T F 1. Electricity is the only mechanical system used in a house.

T F 2. Electric current results when a conductor is passed through a magnetic field.

T F 3. Notches for plumbing should be square or rectangular and then reinforced with metal strapping.

T F 4. A plumber may choose to follow any of the five plumbing codes.

T F 5. It is not possible to operate furnaces without chimneys.

T F 6. The lower the EER rating of an appliance, the greater its efficiency.

Multiple Choice

Choose the answer that correctly completes the statement. Write the corresponding letter in the space provided.

_____ 7. The purpose of an electrical box is to _____.

   A. hold devices such as switches and convenience outlets
   B. enclose connections between conductors
   C. protect connections between two or more conductors
   D. All of the above.

_____ 8. Venting is a part of a plumbing system that _____.

   A. permits air to enter the drain/waste pipes
   B. prevents introduction of wastewater into the water supply system
   C. prevents back pressure and siphoning of water from traps
   D. All of the above.

_____ 9. A _____ heating system heats a building with a boiler and pipes buried in the floor.

   A. forced air
   B. hydronic perimeter
   C. hydronic radiant
   D. resistance radiant

Completion

Place the answer that correctly completes the statement in the space provided.

_____ 10. _____ are the wires that carry electric current from one part of a house-wiring system to another.

_____ 11. A hydronic _____ heating system is one that uses water as a medium for moving the heat from the unit.

_____ 12. _____ is a ropelike, tarred material used to seal lead joints in DWV systems.

_____ 13. Copper tubing comes in rolls and must be unrolled carefully to prevent _____ .

_____ 14. Because they work at such high temperatures, _____ are a source of great energy waste.

_____ 15. _____ cover ends of forced air ducts and distribute warmed air into rooms.

Matching

Select the correct answer from the list on the right and place the corresponding letter in the blank on the left.

_____ 16. Transformer.

_____ 17. Coil.

_____ 18. Direct current.

_____ 19. Alternating current.

_____ 20. Induction.

A. Process by which a magnetic field from one coil causes a current in a nearby coil.
B. Device that can increase or decrease electrical voltage.
C. Electrons flow to and fro in a conductor.
D. Electrons flow in only one direction.
E. Several wrappings of a conductor around an iron bar

_____ 21. Compression fitting.

_____ 22. CPVC.

_____ 23. ABS.

_____ 24. Neoprene.

A. Inexpensive plastic pipe that resists effects of chemicals.
B. Used to make connections in copper tubing.
C. Used as gasket material to seal plumbing joints.
D. A buff-colored thermoplastic used for hot water piping.

_____ 25. Cooling coil.

_____ 26. Condenser.

_____ 27. Thermostat.

_____ 28. Cut-in point.

A. Signals air conditioner to provide cooled air.
B. Coil in a furnace receiving cooled refrigerant from an air conditioner.
C. Exhausts heat from heated refrigerant to the air.
D. Temperature setting that causes a thermostat to call for heat or cooled air.

Identification

Identify the parts of the transformer pictured to the right.

_____ 29. Primary coil.

_____ 30. Secondary coil.

_____ 31. Core.

Name _____

Identify the parts of the air conditioning system pictured below.

_____ 32. Fan.

_____ 33. Condenser.

_____ 34. Register.

_____ 35. Cooling coil.

_____ 36. Blower.

_____ 37. Furnace.

_____ 38. Filter.

_____ 39. Compressor.

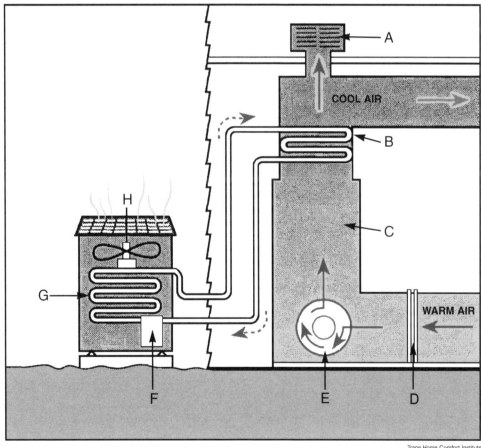

29

Scaffolds and Ladders

OBJECTIVES

Students will be able to:
- ❏ Explain typical designs and construction of manufactured and site-built scaffold.
- ❏ Discuss the types and uses of brackets and jacks.
- ❏ List ladder types and maintenance techniques.
- ❏ Apply ladder and scaffolding safety rules.

INSTRUCTIONAL MATERIALS

Text: Pages 719-728

    Important Terms, page 727

    Test Your Knowledge, page 727

    Outside Assignments, page 727

Workbook: Pages 189-192

Instructor's Manual:

    Reproducible Master 29-1, *Scaffold Designs, Parts A* and *B*

    Reproducible Master 29-2, *Modern Scaffold Assembly*

    Unit 29 Quiz

TRADE-RELATED MATH

The feet of a ladder should be placed away from the building one-fourth the distance to the top support. In other words, if the distance from the ground to the eaves trough is 20′, the feet of the ladder should be placed 5′ away from the building to assure adequate support.

$$20 \div 4 = 5$$

INSTRUCTIONAL CONCEPTS AND STUDENT LEARNING EXPERIENCES

TYPES OF SCAFFOLDING

1. Identify the designs of wooden scaffolds—double-pole and single-pole. Using Parts *A* and *B* of Reproducible Master 29-1, *Scaffold Designs,* discuss the construction of both designs. Stress the importance of selecting straight-grain lumber that is free of large knots. Also mention that the edge grain of the lumber should be parallel to the surface.

2. Identify the components of a scaffold assembly using Reproducible Master 29-2, *Modern Scaffold Assembly.* Stress the importance of setting up the scaffold on a solid base plate, ensuring that the legs will not sink into the ground.

3. Discuss the advantages of using wall brackets when it is necessary to reach high elevations. Emphasize the importance of securely attaching the wall brackets to walls using plenty of 16d or 20d nails, or by bolting it directly to the wall.

4. Describe the use of roof brackets when working on steep slopes.

5. Explain the use of ladder jacks to support simple scaffolds for repair projects. Stress that a lifeline should be used when moving the scaffold, as well as when seated and working on the scaffold.

6. Discuss the use of trestle jacks for interior work. Stress the importance of using good-quality, sound stock for ledger and platform.

7. Review the safety rules for scaffolding found on pages 722 and 723 of the text. Emphasize the importance of adhering to the safety rules when working on the job.

LADDERS

1. Have the students identify the types of ladders commonly used in the construction trade—safety rolling ladders, stepladders, extension ladders, and one-piece straight ladders.

2. Review the safety rules for ladders found on pages 725 and 726 of the text. Emphasize the importance of adhering to the safety rules when working on the job.

3. Discuss the proper means of maintaining a ladder for safe use.

4. Demonstrate the correct, safe procedure for erecting one-piece ladders and extension ladders.

UNIT REVIEW

1. Review the unit objectives. Be sure that the students fully understand each objective.
2. Assign *Important Terms, Test Your Knowledge* questions, and *Outside Assignments* on page 727 of the text. Review the answers in class.
3. Assign pages 189-192 of the workbook. Review the answers in class.

EVALUATION

1. Use Unit 29 Quiz for in-class evaluation. Correct the quizzes and return them to the students for review.

ANSWERS TO TEST YOUR KNOWLEDGE TEXT PAGE 727

1. By rotating adjustable feet on the scaffolding.
2. It can be moved from place to place without disassembly.
3. 2×6
4. 8
5. 8
6. four
7. 8, 26
8. rails
9. 3

ANSWERS TO WORKBOOK QUESTIONS PAGES 189-192

1. A. materials
   B. safety
2. reaching
3. A. Guard rail.
   B. Pole.
   C. Ribbon.
   D. Brace.
   E. Ledger.
   F. Brace.
   G. Blocking/pad.
4. C. 2×6
5. B. 10′
6. diagonal braces
7. A. Roofing bracket.
   B. Ladder jack.
   C. Wall bracket.
8. C. four
9. trestle jacks
10. A. Extension.
    B. Rung.
    C. Rail.
11. C. 25%
12. B. 3′
13. A. Varnish.
14. C. equip bottom end of rails with safety shoes
15. B. At least three legs should rest on a solid support.
16. electrical current

ANSWERS TO UNIT 29 QUIZ

1. True.
2. True.
3. False.
4. False.
5. D. 2×10
6. C. Swinging
7. B. 12″
8. C. one-fourth
9. B. 3′

Reproducible Master 29-1 (Part A)
Scaffold Designs

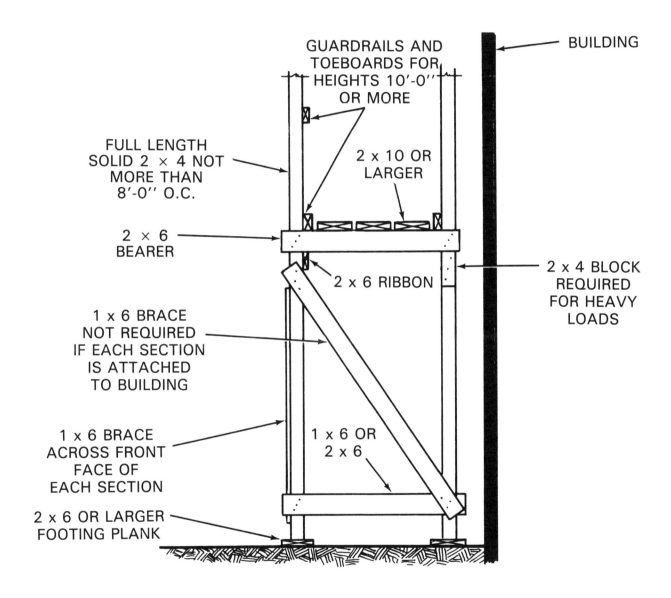

GUARDRAILS AND TOEBOARDS FOR HEIGHTS 10'-0'' OR MORE

BUILDING

FULL LENGTH SOLID 2 × 4 NOT MORE THAN 8'-0'' O.C.

2 x 10 OR LARGER

2 × 6 BEARER

2 x 6 RIBBON

2 x 4 BLOCK REQUIRED FOR HEAVY LOADS

1 x 6 BRACE NOT REQUIRED IF EACH SECTION IS ATTACHED TO BUILDING

1 x 6 BRACE ACROSS FRONT FACE OF EACH SECTION

1 x 6 OR 2 x 6

2 x 6 OR LARGER FOOTING PLANK

Reproducible Master 29-1 (Part B)
Scaffold Designs

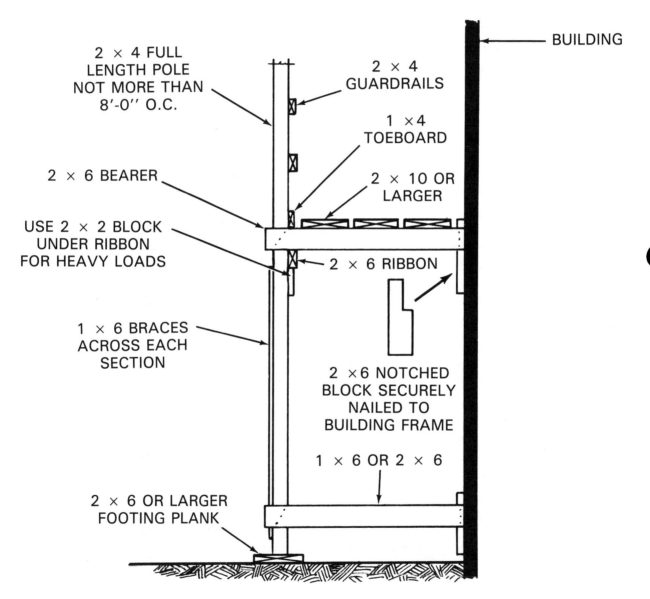

2 × 4 FULL
LENGTH POLE
NOT MORE THAN
8'-0'' O.C.

2 × 4
GUARDRAILS

1 ×4
TOEBOARD

2 × 6 BEARER

2 × 10 OR
LARGER

USE 2 × 2 BLOCK
UNDER RIBBON
FOR HEAVY LOADS

2 × 6 RIBBON

BUILDING

1 × 6 BRACES
ACROSS EACH
SECTION

2 ×6 NOTCHED
BLOCK SECURELY
NAILED TO
BUILDING FRAME

1 × 6 OR 2 × 6

2 × 6 OR LARGER
FOOTING PLANK

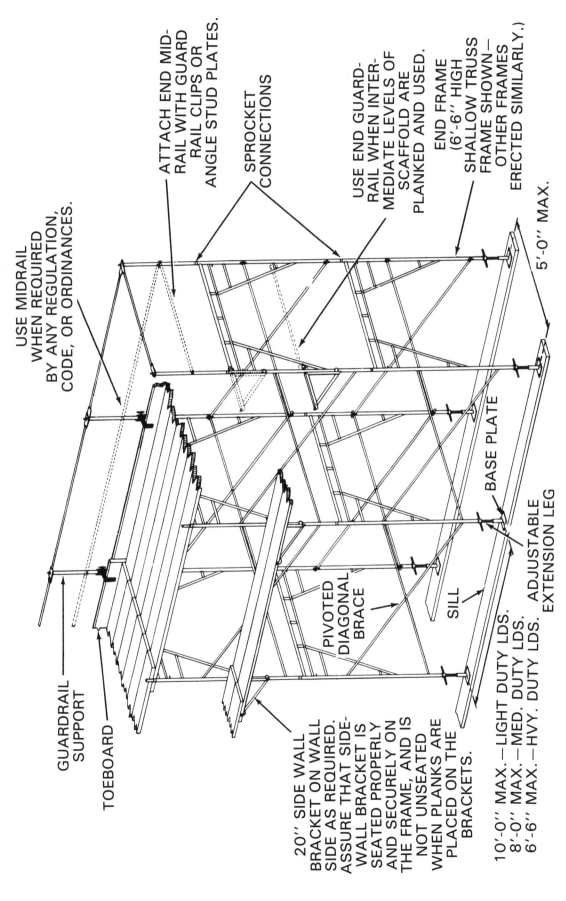

Copyright Goodheart-Willcox Co., Inc.

Reproducible Master 29-2

Modern Scaffold Assembly

USE MIDRAIL
WHEN REQUIRED
BY ANY REGULATION,
CODE, OR ORDINANCES.

ATTACH END MID-
RAIL WITH GUARD
RAIL CLIPS OR
ANGLE STUD PLATES.

SPROCKET
CONNECTIONS

USE END GUARD-
RAIL WHEN INTER-
MEDIATE LEVELS OF
SCAFFOLD ARE
PLANKED AND USED.

END FRAME
(6'-6'' HIGH
SHALLOW TRUSS
FRAME SHOWN—
OTHER FRAMES
ERECTED SIMILARLY.)

5'-0'' MAX.

BASE PLATE

ADJUSTABLE
EXTENSION LEG

SILL

PIVOTED
DIAGONAL
BRACE

GUARDRAIL
SUPPORT

TOEBOARD

20'' SIDE WALL
BRACKET ON WALL
SIDE AS REQUIRED.
ASSURE THAT SIDE-
WALL BRACKET IS
SEATED PROPERLY
AND SECURELY ON
THE FRAME, AND IS
NOT UNSEATED
WHEN PLANKS ARE
PLACED ON THE
BRACKETS.

10'-0'' MAX. – LIGHT DUTY LDS.
8'-0'' MAX. – MED. DUTY LDS.
6'-6'' MAX. – HVY. DUTY LDS.

375

Unit 29 Quiz
Scaffolds and Ladders

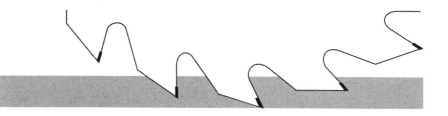

Name _____ Score _____

True-False

Circle T if the answer is True or F if the answer is False.

T   F   1. Some manufactured scaffolding is adjustable as to height of the platform.

T   F   2. For wooden scaffolds, choose lumber in which the edge grain runs parallel to the surface.

T   F   3. Ladder rungs should be painted with a high-visibility color.

T   F   4. Safety shoes must be used for ladders on all types of bearing surfaces.

Multiple Choice

Choose the answer that correctly completes the statement. Write the corresponding letter in the space provided.

_____ 5. Planking for wooden scaffolds must be _____ or larger.

      A. 1×6
      B. 2×4
      C. 2×6
      D. 2×10

_____ 6. _____ scaffolds are suspended from the roof or other overhead structure.

      A. Double-pole
      B. Single-pole
      C. Swinging
      D. None of the above.

_____ 7. Scaffold planking should be lapped _____ and extend 6″ beyond all supports.

      A. 6″
      B. 12″
      C. 18″
      D. 24″

_____ 8. Place a ladder so the horizontal distance from the lower end to the vertical wall is _____ the length of the ladder.

      A. one-half
      B. one-third
      C. one-fourth
      D. one-eighth

_____ 9. A 36′ extension ladder should lap at least _____ .

      A. 2′
      B. 3′
      C. 4′
      D. 6′

30

Carpentry—A Career Path

OBJECTIVES

Students will be able to:

❑ Cite extent of demand for carpenters in coming years as projected by the U.S. Department of Labor

❑ List job possibilities for the trained carpenter.

❑ Describe the sequence of carpentry training and apprenticeship.

❑ Discuss abilities and characteristics needed by those in the carpentry field.

INSTRUCTIONAL MATERIALS

Text: Pages 729-738
  Important Terms, page 737
  Test Your Knowledge, page 737
  Outside Assignments, page 737
Workbook: Pages 193-195
Instructor's Manual:
  Reproducible Master, 30-1, *Apprenticeship Training's Relationship to Related Groups*
  Unit 30 Quiz

INSTRUCTIONAL CONCEPTS AND STUDENT LEARNING EXPERIENCES

ECONOMIC OUTLOOK

1. Using a current edition of the "Occupational Outlook Handbook," prepare a handout for students on the economic outlook for construction.

2. Emphasize that a share of the economic growth is in construction of buildings.

3. Note that there are other areas of construction that also offer opportunities for carpentry skills.

EMPLOYMENT OUTLOOK

1. Discuss size of the current workforce in carpentry (an estimated 990,000). Note that the demand for carpenters is the highest of any of the crafts.

2. Have students look at the cyclic layoffs as well.

3. Suggest that for those interested in a carpentry career, careful budgeting of finances is necessary to bridge the layoffs.

OPPORTUNITIES

1. Discuss the types of job opportunities available for a carpenter.

2. Have the students list factors that should be considered when selecting the type of work they would like to do.

TRAINING

1. List the types of courses that students should take in high school to enhance their career in carpentry. Be prepared to explain why courses such as English or social studies are important in their training.

2. Describe the "roads" that can be taken after graduating from high school to prepare for a career in the carpentry profession.

APPRENTICESHIP

1. Discuss the historic background of the apprenticeship training program and how it has developed into today's program.

APPRENTICESHIP STAGES

1. Describe the basic requirements for a carpentry apprentice.

2. Explain the stages that a person must typically go through when involved in an apprenticeship program. Discuss such things as responsibilities, courses they will be required to take, and wages.

3. Using Reproducible Master 29-1, *Apprenticeship Training and Related Groups*, explain that industry and unions are supportive of apprenticeship programs and offer important learning opportunities for apprenticed carpenters.

PERSONAL QUALIFICATIONS

1. Have the students list the personal qualifications necessary to become a successful carpenter. Include not only work-related traits, but also character traits and interpersonal skills.

UNIT REVIEW

1. Review the unit objectives. Be sure that the students fully understand each objective.
2. Assign *Important Terms, Test Your Knowledge* questions, and *Outside Assignments* on page 737 of the text.
3. Assign pages 193-195 of the workbook. Review the answers in class.

EVALUATION

1. Use Unit 30 Quiz for in-class evaluation. Correct the quizzes and return them to the students for review.
2. Use the Section 7 Exam to evaluate the student's mastery of information found in Units 29-30 of the text.

ANSWERS TO TEST YOUR KNOWLEDGE TEXT PAGE 737

1. West, South
2. D. 990,000
3. Cyclical layoffs.
4. ten
5. Evaluate individually.
6. organizes and starts a business

ANSWERS TO WORKBOOK QUESTIONS PAGES 193-195

1. 100,000
2. C. Heating, plumbing, and electrical maintenance in commercial buildings.
3. drafting
4. D. All the above areas.

5. C. child-parent
6. D. 7 years
7. journeyman
8. management
9. C. Associated General Contractors of America, Inc.
10. B. 17 years
11. C. 4 years
12. D. 144 hr.
13. C. 50%
14. journeyman

ANSWERS TO UNIT 30 QUIZ

1. True.
2. False.
3. False.
4. False.
5. B. 3
6. C. journeyman
7. B. 4

ANSWERS TO SECTION 7 EXAM

1. False.
2. False.
3. True.
4. False.
5. False.
6. True.
7. C. rails
8. B. 3'
9. D. 12"
10. C. 4
11. C. 4'
12. D. 50%
13. F
14. A
15. E
16. D
17. C
18. B

Reproducible Master 30-1
Apprenticeship Training and Related Groups

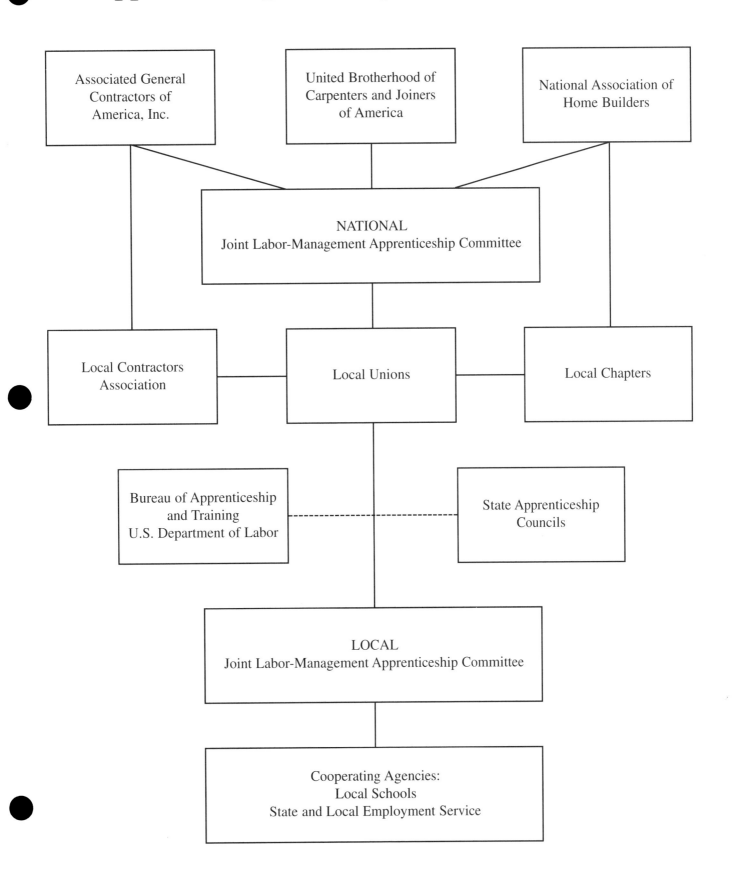

Unit 30 Quiz
Carpentry—A Career Path

Name _____ Score _____

True-False

Circle T if the answer is True or F if the answer is False.

T   F   1. Interpersonal skills are important to being a successful carpenter.

T   F   2. Math and science have little impact on the carpentry trade, and are therefore not important aspects of training.

T   F   3. Apprenticeship training is a recently developed concept.

T   F   4. In modern apprenticeship programs, apprentices are generally not paid.

Multiple Choice

Choose the answer that correctly completes the statement. Write the corresponding letter in the space provided.

_____   5. Nearly one out of _____ carpenters is self-employed.
   A. 2
   B. 3
   C. 4
   D. 5

_____   6. A(n) _____ is a person who is fully qualified in a trade.
   A. apprentice
   B. laborer
   C. journeyman
   D. None of the above.

_____   7. The term of an apprentice carpenter is _____ years under normal conditions.
   A. 2
   B. 4
   C. 6
   D. None of the above.

Section 7 Exam
Scaffolds and Careers

Name _____ Score _____

True-False

Circle T if the answer is True or F if the answer is False.

T F   1. Guardrails and toeboards are required for scaffolds only at heights greater than 25'.

T F   2. Carpenter apprentices must be at least 25 years old.

T F   3. The term of a carpentry apprenticeship is four years.

T F   4. Courses such as science and math are not really necessary to understand the technical aspects of modern plumbing.

T F   5. A journeyman is a person learning a trade or craft.

T F   6. Apprenticeship training consists of on-the-job, as well as classroom training.

Multiple Choice

Choose the answer that correctly completes the statement. Write the corresponding letter in the space provided.

_____ 7. The sides of a ladder are called _____.

   A. rungs
   B. stiles
   C. rails
   D. None of the above.

_____ 8. When a ladder is used to climb onto a roof, it should extend at least _____ above the roof edge.

   A. 2'
   B. 3'
   C. 4'
   D. 5'

_____ 9. Scaffold planking should lap at least _____ .

   A. 3"
   B. 6"
   C. 9"
   D. 12"

_____ 10. The height of a scaffold platform should not exceed _____ times the smallest base dimension.

   A. 2
   B. 3
   C. 4
   D. 5

_____ 11. A 48′ ladder should lap at least _____ between sections.
   A. 2′
   B. 3′
   C. 4′
   D. 5′

_____ 12. The initial pay scale for an carpenter apprentice is about _____ of a skilled worker.
   A. 15%
   B. 25%
   C. 40%
   D. 50%

Identification

Identify the parts of the double-pole scaffold.

_____ 13. Footing plank.

_____ 14. Guardrail.

_____ 15. Brace.

_____ 16. Ribbon.

_____ 17. Toeboard.

_____ 18. Bearers.